AF718238

Protection of workers against radio-frequency and microwave radiation: A technical review

OCCUPATIONAL SAFETY
AND HEALTH SERIES
No. 57

PROTECTION OF WORKERS AGAINST RADIO-FREQUENCY AND MICROWAVE RADIATION: A TECHNICAL REVIEW

INTERNATIONAL LABOUR OFFICE GENEVA

ISBN 92-2-105604-X
ISSN 0078-3129

First published 1986

Printed in the German Democratic Republic ZIM

PREFACE

This review aims to provide basic guidance on working conditions and procedures that will lead to higher standards of safety for all personnel engaged in the manufacture, maintenance and operation of radio-frequency (RF) and microwave (MW) devices. The following areas are covered: the characteristics of RF and MW radiation; biological effects and effects on humans; occupationally related exposure type and effects, and community effects; instrumentation and measuring techniques; occupational exposure limits; control of and protection from exposure to RF and MW radiation; and the administrative structure for the management of non-ionising radiation.

As an additional basis, this document draws on ILO Occupational Safety and Health Series No. 53, entitled *Occupational hazards from non-ionising electromagnetic radiation*, which was prepared in collaboration with the International Non-Ionising Radiation Committee (INIRC) of the International Radiation Protection Association (IRPA). Collaboration with IRPA/INIRC continued in preparing a manuscript as a basis of this publication. This publication is one of a series of studies on occupational hazards arising from non-ionising radiation, being carried out as part of the International Programme for the Improvement of Working Conditions and Environment (PIACT).

The Office expresses its gratitude to IRPA and INIRC for their continuing collaboration in promoting occupational safety and health.

PREFACE

This manual [illegible] to provide basic guidance [illegible] and [illegible] that will lead to higher standards of safety for all personnel engaged in the [illegible] and operation of [illegible] are covered. The characteristics of RF and the [illegible] effects and [illegible] standards and [illegible] techniques [illegible] RF [illegible] for the management of [illegible]

As a [illegible] this document [illegible] Safety and Health [illegible] the [illegible] International Occupational [illegible] (INIRC) of the International Radiation Protection Association (IRPA). Collaboration with IRPA/INIRC continues [illegible] of [illegible] working conditions and [illegible] (PIACT).

The Office expresses its [illegible] for [illegible] occupational [illegible]

CONTENTS

Page

Page

.../...

1. Introduction

Radio-frequency workers may be defined as persons in the operation, manufacture or maintenance of radio-frequency radiation-emitting devices. It should be understood that people taking medication, pregnant women and persons with various disabilities are included within this definition, and so account must be taken of this when preparing safety guide-lines.

Concerns about radio-frequency induced effects and possible health hazards first arose in industrialised countries, but now the problem has attained a universal character. The number of devices, such as radars, broadcast towers and industrial equipment which emit RF electromagnetic waves, is growing from day to day all over the world.

Surveys of radio-frequency equipment used in industry suggest that one of the greatest sources of occupational exposure comes from industrial heating devices, predominantly dielectric (RF) heaters. RF induction heaters and furnaces may also fall into this category, although, little confirmed data are available on the exposure levels from these devices. Broadcast towers or radar installations where maintenance personnel must either climb close to the antennae or feed cables, or repair equipment while it is operating, also constitute a potential source of excessive radiation exposure. Hand-held communication devices where human exposure occurs close to radiating antennae in the near field may present special concerns.

2. Characteristics of radio-frequency (RF) and microwave (MW) radiation

For the purpose of this review, RF radiations are defined as occupying that part of the electromagnetic spectrum having frequencies from 100 kHz to 300 GHz or wavelengths in space from 3 km to 1 mm. Included within this range are MW radiations having frequencies from 300 MHz to 300 GHz or wavelengths from 1 m to 1 mm (refer to the spectrum of electromagnetic radiation in figure 2.1).

As the wavelength depends on the dielectric properties of the medium through which the wave propagates, it is better to characterise MW and RF radiations by their frequency, since this does not change with the medium. It may be noted here that the wavelengths of radiations in this range are much smaller in biological media than in air.

Microwave bands may also be designated by letters or by metric wavelengths (5). In some American publications (62) microwaves are designated as waves of frequencies between 10 MHz and 100 GHz. Various designations used for particular regions of the RF radiation spectrum are presented in table 2.1.

Figure 2.1: The spectrum of electromagnetic radiation

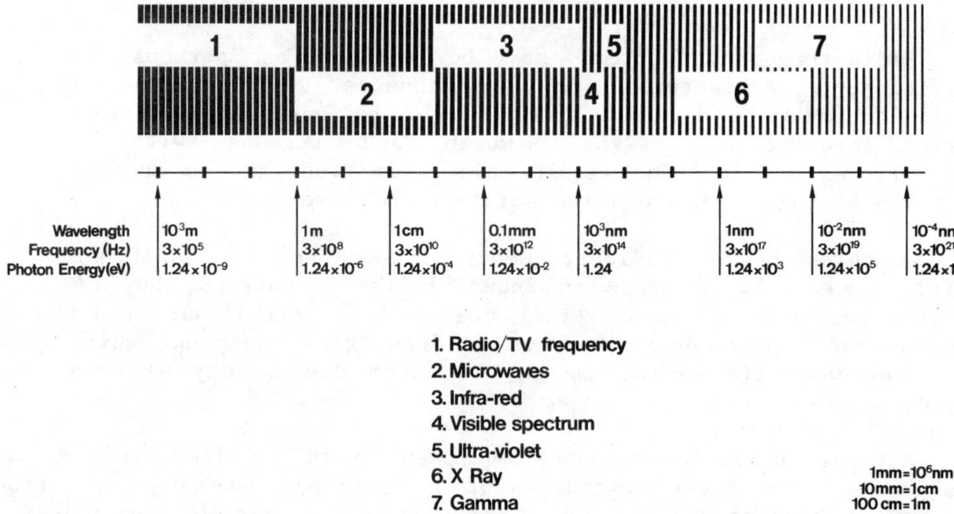

2.1 Physical properties

Electromagnetic waves are composed of electric and magnetic fields, propagated as periodic disturbances in a vacuum or in a physical medium. These waves propagate through free space with the velocity of light (3×10^8 m/s), with transverse oscillations perpendicular to the direction of propagation. For a plane wave the directions of the electric and magnetic fields are perpendicular to each other and perpendicular to the direction of propagation as shown in Appendix 2, figure 1,. The fields (both electric and magnetic) oscillate in time from their maximum positive value (E_o or H_o) through zero to their maximum negative value. The frequency of oscillation can be defined as the number of cyclic changes per second (e.g. the number of maximum positive values in one second). The wavelength (shown in Appendix 2, figure 1) is the distance between the two consecutive maxima (or minima). The relationship between the frequency and wavelength is $f = c/\lambda$, where f is the frequency, c is the velocity of light and λ is the wavelength.

For further details on the physical properties of radiation, refer to Appendix 2.

2.2 Units of radiation

RF electromagnetic radiation induces conduction current in conductors and displacement current in semiconductors (imperfect dielectrics). These currents, especially displacement current in semiconductors, are responsible for the transformation of electromagnetic energy into heat. Thus the radiation intensity can be determined calorimetrically. In SI terminology, it is known as the irradiance and is expressed in W/m^2. However, the unit has been frequently reported in mW/cm^2 and $\mu W/cm^2$, especially in conjunction with the term "power density".

Table 2.1: Radio-frequency and microwaves designations

Aproximate wavelength in air (m)	Frequency (MHz)	Designations				
		Used traditionally	Russian literature	Metric, used USSR, Poland and Czechoslovakia	English and American literature	Loosely used
15.10^3 and more	0.2 and less					
10^4 --	-- 0.03					
		Very long waves	NTch, niskie, tchastoty (low frequency)	Decamyria and myriametre waves	ELF, extremely low frequency	AF, audio frequency
3.10^3 --	-- 0.1					
1 000 --	-- 0.3				LF, low frequency	
800 --	-- 0.4					
100 --	-- 3					
		Long waves	Srednyie, tchastoty (median frequency)	Kilometre waves High frequency Hectrometre waves	MR, median frequency	RF radio-frequency Hertzian waves
30 --	-- 10	Median waves			HF, high frequency	
10 --	-- 30	Short waves		Decametre waves		
1.5 --	-- 200					
		Ultra-short waves	UV Tch, ultravysokie tchastoty (ultra-high frequency)	Metre waves		
1 --	-- 300					
10^{-1} --	-- 3 000	Microwaves	SVtch, sverch wysokie tchastoty (super-high frequency)	Decimetre waves	UHF, ultra-high frequency	Designated as micro-waves, radar emanations, UHF or HF
10^{-2} --	-- 30 000					
$1.5.10^{-3}$ --	-- 200 000			Centimetre waves	SHF, super-high frequency	
10^{-3} --	-- 300 000					
				Millimetre waves	EHF, extra-high frequency	
$0.75.10^{-3}$ --	-- 400 000	Intermediate and IR (infrared) radiation	IK, Infra krasnye (infrared)	IR (infrared)	IR (infrared)	IR (infrared)
$0.3.10^{-3}$ --	-- 1 000 000					

Source: Baranski and Czeiski (5).

The principal equations of electromagnetism, formulated by Maxwell, are based on intrinsic symmetry between the electric field $\vec{E}$, magnetic field $\vec{H}$ vectors and propagation vector $\vec{K}$ of an electromagnetic wave. In the region of RF electromagnetic radiation, two zones are distinguished, namely the near field (or Fresnel zone) close to source, and the far field (or Faunhofer zone) far from source. The radius denoting the border between the near and remote fields may be calculated approximately from the formula $R = 2D^2/\lambda$, where D denotes larger dimension of antenna and λ denotes wavelength. The electric-field and magnetic-field vectors are perpendicular to each other at every point of an electromagnetic wave far from its source and both these vectors are perpendicular to the direction of propagation vector $\vec{K}$. Such a field of electromagnetic radiation is of plane-wave polarisation characterised by the ratio E/H equal to 377 ohms as in free space. The amount of energy conveyed by an electromagnetic wave is proportional to the product of the amplitudes of the electric and magnetic vectors or the square of the amplitude of either of these two vectors. Interaction of this energy with matter is accompanied by its transmission, diffraction or absorption. In the near field of radiation and around electromagnetic scattering objects, the vectors $\vec{E}$, $\vec{H}$ and $\vec{K}$ may have different configurations and different relative magnitudes. Whereas in the near field the electromagnetic energy can be measured in V/m for the electric field and in A/m for the magnetic field, in the far field in the case of plane-wave radiation it is possible to measure the field strength E in V/m, the field strength H in A/m and the "power density" P in W/m^2 since these parameters are related by the Poynting vector $\vec{S} = \vec{E} \times \vec{H}$ where S by simplication denotes energy-flow density in W/m^2 (32).

In far-field conditions, values for the power density can be obtained from E and H as shown in table 2.2.

Table 2.2: Power densities in the more frequently reported units for free-space far-field conditions

W/m^2	mW/cm^2	$\mu W/cm^2$	V/m	A/m
10^{-2}	10^{-3}	1	2	5×10^{-3}
10^{-1}	10^{-2}	10	6	1.5×10^{-2}
1	10^{-1}	10^2	2×10	5×10^{-2}
10	1	10^3	6×10	1.5×10^{-1}
10^2	10	10^4	2×10^2	5×10^{-1}
10^3	10^2	10^5	6×10^2	1.5
10^4	10^3	10^6	2×10^3	5

3. Effects of RF and MW

A complete review of the scientific literature on the biological effects of exposure to RF and MW radiation is beyond the scope of this document. However, a number of good reviews are available (5, 10, 33, 39, 63).

The following is a brief review of the biological effects on experimental animals, as well as the health effects observed in humans.

3.1 Biological effects of RF and MW

Selected reports in the scientific literature have been evaluated in terms of whether or not the biological effects constituted a hazard to health. Behaviour in experimental animals seems to be the most sensitive indicator of an adverse health effect (e.g. convulsion activity, work stoppage, work decrement, decreased endurance, perception of the exposing field and aversion behaviour). Some of these effects were demonstrated in experimental animals exposed to power densities as low as 5 mW/cm^2. On the basis of their reviews, the Bio-effects Subcommittee of the United States American National Standards Institute - ANSI (59) concluded that exposure to RF electromagnetic energy that is deposited in the whole body at an average specific absorption rate (SAR) of less than 4 W/kg does not produce an adverse health effect in experimental animals.

3.1.1 Energy absorption in biological systems

Many observed biological effects seem to depend on rates of energy deposition. For example, relationships have been demonstrated between the energy absorbed and temperature elevation. These relationships depend on the thermoregulatory and adaptive mechanisms under various environmental conditions. These modify the relationship between absorbed energy and the generalised rise in body temperature. Moreover, the energy deposition may be quite non-uniform and significant localised temperature gradients can arise, making the prediction of effects difficult. In fact, such gradients may be important in contributing to the biological response.

In view of the differences between empirical and theoretical observations, the significance of "non-thermal" effects is indicated. This is also hypothesised by the direct interference with bioelectric phenomena and the role of electromagnetic fields in the transmission of biological information.

For frequencies below 10 MHz very little information on biological effects exists, except in the related areas of bone growth and wound healing. The rate of energy absorption by the human body rapidly decreases at lower frequencies. However, because one cannot necessarily assume that a biological effect is directly related to the amount of energy absorbed, greater safety margins should be incorporated into the exposure limits than would be suggested from the whole-body-averaged SAR for humans. The ANSI Committee (59), for example, chose a whole-body, time-averaged upper limit of 0.4 W/kg, with a peak SAR not to exceed 8 W/kg when averaged over a time of 6 min for 1 g of tissue.

3.1.2 Molecular energy absorption and effects

The amount of energy per photon of RF radiation is too small to cause ionisation. Such an energy level is sufficient, however, to excite the vibration of macromolecules, molecules and atoms, and, as already mentioned, polarisation of molecules and atoms may also occur. RF radiation exerts a catalytic action upon some chemical and enzymatic reactions: this effect is often limited to a very narrow frequency band.

The reproducible experimental results indicate that selective extremely low frequency fields at 6 to 12 Hz of as low intensity as 0.1 to 0.5 V/m caused a decrease by 12 to 15 per cent of calcium efflux from the perfused animal brain. Moreover, amplitude or frequency modulation within 6 to 20 Hz of an electromagnetic carrier wave in the range 147-450 MHz and with a field intensity between 0.1 and 1.0 mW/cm^2 induced a 15 per cent increase of calcium efflux from the peripheral animal brain. Other modulatory frequencies and field intensities had no effect on calcium efflux. The above-mentioned results strongly suggest that within the spectrum of RF electromagnetic radiation there are certain amplitudes and frequencies that are biologically effective in the binding and release of calcium at the surface of cell membranes (32).

Where higher frequencies are involved, namely in the region of 60-120 GHz, studies have suggested that alteration of cell division and virus inactivation or activation in macromolecules takes place in biological systems (23, 46).

The possibility of such effects also include enzyme systems, chromosomes and cell membranes (25).

3.1.3 Effects on experimental animals

Most of the data collected on the effects of RF and MW radiation comes from research on experimental animals. The following is a brief summary of such effects.

3.1.3.1 Thermal and related effects

Thermal effects have been attributed to the conversion into heat of absorbed MW and RF radiation. Injuries included local lesions as well as gross thermal stress due to hyperthermia. Death from hyperthermia was related to power densities of few mW/cm^2 to several thousands mW/cm^2, radiation frequency and animal size. Lesions have also been found in internal organs of animals exposed to radiation for prolonged durations.

3.1.3.2 Effects on the eye

The eye is provided with a limited blood supply, so that heat deposited in this tissue is carried away slowly. Because of this, high-power-density MW radiation causes sufficient heating of the eye to result, in some cases, in irreversible eye damage such as lens opacities, cataracts, corneal damage and retinal lesions. Reversible irritation and inflammation of the conjunctiva, cornea and iris can also occur depending on the intensity of the incident radiation. The threshold level for any irreversible changes from a single long exposure (i.e. greater than 30 min) depends on frequency. For frequences less than 2 GHz, the first injuries to appear are increased lens opacities and cataracts, with a threshold estimated from animal studies to be at least 100 mW/cm^2 for one hour of continuous exposure. At frequencies above 10 GHz the first permanent eye damage is to the cornea, and the single-exposure threshold power density for 35 GHz and higher frequencies could be as low as 30 mW/cm^2 for a one-hour exposure.

3.1.3.3 Nervous, haematopeitic and immunological effects

Exposure to a 10 mW/Cm^2 demonstrated behavioural, bioelectrical, metabolic and structural changes in rodents. Immunological reactivity changes caused nervous system changes which were influenced by the susceptibility of animals and other environmental factors.

3.1.3.4 Reproductive and genetic effects

Chromosomal aberrations and disturbances in somatic cell divisions, germ cell maturation and spermatogenesis have been reported as genetic effects. Although no threshold levels or dose-effect relationships have been established, a thermal-induced type of mechanism is implicated.

3.1.3.5 Central nervous system (CNS), cardiovascular function and thermoregulation

Experimentation with laboratory animals in the Soviet Union is reported to demonstrate some clinical effects arising from exposure to RF and MW radiation (43). Electroencephalogram (EEG) measurements, mainly in rabbits, have been presented as evidence for a direct influence of microwaves on brain tissue independent of heating influences. An increase in the presence of slow-wave EEG activity (synchronisation) is said to occur when the head of an animal is irradiated, and is thought to represent an inhibitory effect of microwaves on brain processes. Irradiation of body regions other than the head may induce an alert EEG pattern (desynchronisation). Most of the animal EEG studies are unreliable, since metal recording electrodes were used, and their influence in distorting and concentrating the electromagnetic fields and inducing effects through local heating should not be overlooked. To date, no substantial relationship has been established between MW/RF exposure and EEG response.

Much research on the behavioural effects of microwaves has dealt with conditioned reflexes, e.g. conditioned salivation in dogs, since it was suggested that microwave-induced changes in nervous system function can be most distinctly revealed by this approach. At power densities greater than 10 mW/cm^2, both exaggeration and inhibition of conditioned reflexes have been reported. Other types of behavioural experiments have made it clear that animals can detect the presence of microwaves. Rats will avoid the presence of pulsed microwaves used as a stimulus in conditioning experiments, probably because pulsed microwave energy can, under certain conditions, be perceived through the sense of hearing.

Some investigations in humans and laboratory animals suggest that chronic low-level microwave exposure lowers the arterial blood pressure, and that changes in heart rate can occur during acute exposure. Other researchers have reported changes in the rhythm of isolated hearts produced by low-level microwaves (5). Some serious criticisms of technique and interpretation of data can be made of reports which claim to demonstrate non-thermal effects on cardiovascular function, and these effects remain to be confirmed through careful research. One explanation for blood pressure and heart-rate changes, and behavioural arousal during microwave exposure is that heat stimulation of sensory nerve endings beneath the skin may activate neural mechanisms in the brain which, in turn, evoke the release of adrenal hormones that strongly influence the cardiovascular system.

Exposure to microwaves at relatively high incident power densities, usually above 10 mW/cm^2, produces body heating which must be eliminated through thermoregulatory mechanisms. The circulatory system plays an important role in carrying heat deposited in interior regions of the body to exterior surfaces where it is lost to the environment. Body regions which do not have a plentiful blood flow to carry heat away are more susceptible to thermal injury produced by microwaves, e.g. the eye and testes. Heat loss provided by the evaporation of water from the respiratory tract (optimised by panting in certain species) and from the skin (during sweating) is an important supplemental mechanism for the elimination of a heat load.

3.1.4 Conclusions: Dose-effect relationship

For biological end-points such as a lens cataract of the eye, a time-intensity threshold has been established. A minimum exposure to approximately 1.5 kW/m^2 for at least one hour is needed to produce a cataract (63). Above this threshold, a dose-response relationship has been observed. More complex relationships are evident for other biological effects where no apparent dose-response relationship exists, e.g. effects on the immune system or on calcium efflux in cerebral tissue occurring within frequency or power windows.

The scientific literature by no means precludes the possibility that the whole-body average SAR of a 4 W/kg is not a true threshold below which adverse health effects should not occur. Threshold exposure conditions for biological effects applicable to humans exposed to all parts of the frequency range and to all possible modulation frequencies are not known. Thus safety factors must be incorporated into the exposure limits to allow not only for the lack of scientific data but for all possible conditions under which the exposure might occur.

3.2 Health effects of RF and MW in humans

Studies of occupationally exposed human beings have been conducted but suffer from serious limitations (47). Reported personal exposure is unreliable due to the lack of adequate instrumentation. In most studies, exposure conditions were undefined, making it impossible to estimate such essential parameters as the average power density and durations of exposure, effective area of irradiation and position or orientation of the person with respect to the field. Other environmental factors and socio-economic conditions play a significant role in epidemiology, making it difficult to demonstrate a causal relationship between a disease and exposure. Epidemiological studies (47, 51) have not indicated any reliable evidence of detrimental effects associated with occupational exposure to RF and MW radiation. Controlled exposures of humans have been conducted for therapeutic purposes (short-wave and microwave diathermy). However, because of specific exposure conditions, namely partial-body irradiation at high intensities for relatively short periods of time, this situation is not representative of occupational exposure conditions.

However, it is convenient in discussing the effects of RF and MW radiation to classify such effects into thermal and non-thermal. The following is a summary of the health effects of RF and MW radiation in man as observed and documented from workers' exposure, as well as those effects that have a probability of occurrence as far as can be judged from strictly animal experimentation and likely occupational exposure.

3.2.1 Observed effects

Prior to discussing the observed effects of RF and MW radiation in man, emphasis should be placed on the variables involved, as mentioned above, in arriving at the conclusions regarding such effects.

3.2.1.1 CNS, cardiovascular function and thermoregulation

Clinical symptoms attributed to MW or RF radiation have been described from studies in occupationally exposed persons (43). The symptoms are numerous and difficult to classify, involving changes in mental function and other disturbances of body physiology. The causal relationship between exposure and the clinical findings is questionable (4). The microwave power density to which these persons were exposed is often unknown, and some clinicians have suggested that many of the health problems arise from the patient's own personality problems, e.g. hypochondria. The subjective nature of clinical assessment of patient complaints and the poor use of control groups have also obscured the issue.

Human beings, like other animals, are able to perceive heating from microwaves through sensory nerve fibres in the skin. When they anticipate and are attentive to such stimuli under experimental conditions, the threshold power densities that produce thermal sensations are far below those that cause thermal damage and pain. A sensation of warmth at 3 GHz is produced after an exposure of 1 to 4 s duration at incident power densities of 21 mW/cm^2 to 12.6 mW/cm^2. An exposure of 20 s duration at 3,100 mW/cm^2 produces a sensation of pain (27).

3.2.1.2 Effects on the eye

There are five reliable surveys comparing the eyesight of microwave workers with that of a control group. None show significantly more cataracts in the exposed group. Only one of the five showed significantly more subclinical lenticular opacities, the precursor to cataracts. The same survey found significantly more cases of retinal lesions in the exposed group as well.

3.2.1.3 Effects on hearing

Under certain microwave exposure conditions humans perceive an auditory sensation when the head is exposed to rectangular pulse-modulated microwave energy (200-3,000 MHz) with peak incident power densities of the order of 300 mW/cm^2, average power densities as low as 0.1 mW/cm^2 and pulse widths varying from 1 to 100 μs. The auditory sensation is heard as a click, buzz or chirp, depending on the pulse width and pulse-repetition frequency, and is perceived as originating from within or near the back of the head. The absorbed microwave energy produces a thermoelastic expansion of the brain tissue causing an acoustic pressure wave which is detected by the hair cells of the organ of Corti (in the cochlea) via bone conduction. The deposition of energy needed to induce this auditory sensation is very small and has been estimated to increase the temperature of the brain per pulse by less than 10^{-5} °C. The auditory sensation can be loud enough to be clearly perceived by humans in an ambient noise level of 65 dB.

This effect is judged not to be a hazard as it results from the very great sensitivity of the inner ear to the minute thermoacoustic pressure wave created by the pulse (33).

3.2.1.4 Other effects

An epidemiological study by Lin et al. (37), using data from death certificates of 951 adult white males, showed that a significantly higher proportion of primary brain tumours was found among men employed in electricity-related occupations when compared with a control population. An increase in the odds ratio for brain tumours was found to be positively related to electromagnetic field exposure levels.

A study by Logue et al. (38) based on data collected for a larger retrospective mail survey of potential health effects of occupational exposure to RF/MW and other energy on male physiotherapists described the relationship of paternal occupational exposure to non-ionising radiation and congenital anomalies in offspring. However, this study did not show non-ionising radiation to be a possible aetiologic factor.

3.2.2 Probable effects

The probable effects of RF and MW radiation in humans refer to those effects which have been substantiated by animal experimentation only, namely those related to the reproductive and genetic effects.

3.2.3 Conclusion

For the purpose of this document, the discussion on the hygiene aspects of RF and MW radiation, namely the protection from and control of exposure to such radiation (see section 8), shall regard both the observed and probable effects, under sections 3.2.1 and 3.2.2 respectively, as hazardous. The adoption of this strategy aims at securing an additional safety factor for the protection of the workers.

4. Sources of occupational exposure to RF and MW

4.1 Natural sources

RF electromagnetic radiations are emitted naturally into the environment from the sun and galaxies, the earth and from the human body (for a short review see references 40 and 48). However, the natural background of these radiations is small by comparison with the overall intensity of man-made sources (63).

4.2 Man-made sources

Radio-frequency energy is generated by equipment incorporating, for example, oscillator circuits, klystrons, magnetrons and solid-state devices. This energy can be transmitted to an applicator or antenna through a waveguide or transmission line. Waveguides can be rectangular or circular in shape, and may be open, but are normally constructed of a conducting metal such as aluminium, brass or copper, or a non-conducting substance with a thin conducting coat on its inner surface. The waveguide is normally at ground potential.

Radiating antennae such as dipoles, monopoles and parabolic reflectors are typically used to transmit RF energy through free space or through a dielectric material. Transmitted energy can be applied directly to a specialised device, used in communications, or converted to another form of energy, such as heat. Microwave ovens incorporate a waveguide to deliver microwave energy to heat food in a cavity.

The International Telecommunication Union has mediated the rational use of the electromagnetic spectrum by subdividing the part of the spectrum commonly used for telecommunications (radio spectrum) into nine frequency bands, seven of which are in the radio-frequency range as defined in this review (table 4.1). Certain frequencies within these bands have been designated for unlicensed industrial, scientific and medical (ISM) applications. These frequences are shown in table 4.2.

Table 4.1: Nomenclature of frequency band designations used in radiocommunications and some typical applications

Band No.	Band Designation	Frequency range	Wavelength	Metric subdivision	Typical applications
5	LF	30-300 kHz	10-1 km	Kilometric waves	Power line carrier, marine and aeronautical mobile, radio navigation, radio location
6	MF	300-3,000 kHz	1-0.1 km	Hectometric waves	Communications, radio navigation, marine, radio-phone, amateur radio, industrial RF heating equipment, AM broadcast
7	HF	3-30 MHz	100-10 m	Decametric waves	Citizens' band, amateur radio, medical diathermy, AM broadcast, international communications, industrial RF equipment
8	VHF	30-300 MHz	10-1 m	Metric waves	Police, fire, amateur radio, FM, VHF-TV, industrial RF equipment, FM broadcast
9	UHF	300-3,000 MHz	1-0.1 m	Decimetric waves	Short-range broadcast, communications, amateur radio, taxi, police, fire, radar, citizens' band, radio navigation, UHF-TV, microwave ovens, medical diathermy, industrial heating
10	SHF	3-30 GHz	10-1 m	Centimetric waves	Satellite communications, radar, amateur radio, microwave relay, airborne weather radar
11	EHF	30-300 GHz	10-1mm	Millimetric waves	Terrain avoidance, radar, microwave relay

Notes: (1) Band number "N" extends from 0.3×10^N to 3×10^N Hz.

(2) Hz = hertz; k = kilo (10^3); M = mega (10^6); G = giga (10^9).

Source: IRPA/INIRC.

Table 4.2: Frequencies designated for unlicensed use world-wide in industrial, scientific and medical (ISM) applications

ISM band frequencies	
6.78 MHz ± 1.5 kHz[a]	2.45 GHz ± 50 MHz
13.56 MHz ± 6.78 kHz	5.80 GHz ± 75 MHz
27.12 MHz ± 160 kHz	24.125 GHz ± 125 MHz
40.68 MHz ± 20 kHz	61.0 GHz ± 250 MHz[a]
433.92 MHz ± 20 MHz[a,b]	122.5 GHz ± 0.5 GHz[a]
915 MHz ± 13 MHz	245 GHz ± 1 GHz[a]

[a] Special authorisation required for use.

[b] Authorised for use in certain countries only.

Source: World Administrative Radio Conference, Geneva, 1979.

4.2.1 Communications and broadcasting

Although one of the largest applications of RF energy is in communications and broadcasting, there appear to be only a few situations related to these sources where excessive occupational exposure is likely to occur. Broadcast transmitters may constitute a potential health hazard to maintenance crews servicing equipment on the towers while transmitters are operating. Measurements performed in and around television and frequency modulated (FM) stations indicated power densities as high as 180 mW/cm^2 existed on the tower structures, and fields exceeding 1,000 V/m and 3 A/m were recorded immediately in front of television and FM antennae (on the outside of the tower). Inside towers, near FM antennae, electric fields of about 400 V/m and magnetic fields of 2.3 A/m have been measured - predominantly around the cables feeding the antennae (55, 48). Sometimes these cables are located very close to the ladder used to climb the tower and could be responsible for the high exposures received by tower climbers. There is also the possibility that personnel working inside broadcast towers near cables, power dividers, combiners and waveguide flanges that feed antennae could be exposed to high levels of RF radiation.

In some cases it may be necessary to make repairs near an opening in an energised waveguide (hot patching). High exposure levels occur near these openings, depending on the power transmitted. Under these circumstances, special precautions such as using protective gloves or working quickly may be necessary.

4.2.2 Radars

Radars can be divided into high power (acquisition and tracking for civilian and military use) and medium to lower power (on board aircraft and marine craft, and speed control). High-power radars used in conjunction with satellite systems have average effective radiated powers (ERP) up to about 30 GW. Acquisition and tracking radars typically operate at 1-9 GHz and have transmitter powers up to 150 kW. Acquisition radars normally have rotating or nodding antennae which reduce the time-averaged radiation level at a given location. Tracking radars on the other hand, because they have non-rotational antennae and are capable of producing significant radiation levels over large areas, are normally located in remote regions. Near airports or military bases persons may be exposed to sidelobes or secondary radiation from radars with stationary or rotating antennae. Average power density levels can be as high as between 10 and 100 $\mu W/cm^2$ at distances up to 0.8 km from these systems (26).

Low- to medium-power weather radars in aircraft are usually located in the nose under a microwave transparent radome (housing). They typically operate at 4.5 or 9.375 GHz with peak powers of 20-100 kW. Approximate power densities of 15 mW/cm^2 occur at the radome surface to less than 3 mW/cm^2 at a distance of 1 m (56). Military search radars on planes can be much more powerful and would give greater exposure levels.

Radiation from most radars is pulsed, with pulse widths and pulse repetition frequencies that produce an average transmitter power at least two orders of magnitude less than the peak power. In addition, many radar antennae rotate through 360° or oscillate in a sector scan so that the time-averaged power density at a given location will be much less than values obtained for a stationary antenna. Although the time-averaged power densities may be low in the main beam of pulsed radars, caution may be necessary when approaching radars transmitting very high peak pulse values.

4.2.3 Mobile communications devices

Personal communication devices, such as those used by police, firemen and for industrial site control, and in some countries citizen's band (CB) radios mounted on cars or hand-held, such as walkie-talkies, operate at frequencies from 27 to 54 MHz, 132 to 174 MHz and 403 to 512 MHz, generally with a maximum permitted power of 100 W. In and around radio-equipped vehicles, and close to the stub antenna of hand-held radios, substantial fields (up to 1,350 V/m) can occur near the antenna, but these fields fall off rapidly as the distance from the antenna increases (35). It is difficult to interpret these data in terms of human exposure because most measurements are conducted in the near field where the fields are quite complex, so that the absorption in the human body may be quite different. The operation of these devices is intermittent, which also complicates this assessment of partial, non-uniform human exposure.

4.2.4 Microwave ovens

Industrial microwave ovens normally operate at the ISM band frequencies of 915 or 2,450 MHz. Many of these ovens are custom-made for a specific purpose and so are normally surveyed to determine radiation-leakage levels after installation. Surveys of conveyorised industrial microwave heating systems operating at 915 or 2,450 MHz with output powers of 2-60 kW indicated that maximum leakage can occur at the conveyor slot (mostly up to 25 mW/cm^2) (29). However, exposures at the eye level of the operator do not exceed 4 mW/cm^2. In surveys of industrial microwave power devices the maximum potential exposure was found to be approximately 1 mW/cm^2 for all operators under normal working conditions (18).

Most microwave-oven regulations in the world (30, 21, 60, 36) apply only to domestic use. Some countries require that the leakage radiation should not exceed 1 mW/cm^2 at 5 cm from any external surface, at the time of sale, and allow the leakage limit to rise to 5 mW/cm^2 at 5 cm from the outer surface, after the oven has been sold. Only a few of these regulations (e.g. 11) apply to industrial microwave ovens. Personal exposure from microwave ovens is normally small due to a rapid decrease in the power density with increasing distances from the oven. For an oven leaking 5 mW/cm^2 at 5 cm, the power density at 0.3 m is less than 0.15 mW/cm^2 and at 1 m is about 10 $\mu W/cm^2$.

4.2.5 Dielectric (RF) heaters

Dielectric heaters utilise RF energy for heat processing of dielectric materials. Typically these devices generate heat by means of an RF generator and incorporate a power supply, control circuitry, a hydraulic press, a parallel plate applicator (die) and a system for supporting and moving processed material. RF heaters are used to heat, melt or cure materials such as plastic, glue or rubber. Examples of industries involved with these applications are shown in table 4.3.

RF heaters operate within the frequency range 3-70 MHz, with a few (used predominantly for moulding plastics) operating at 300-400 MHz. Most heaters used for sealing operate at frequencies as low as 3-6 MHz. Heaters that are poorly designed or shielded can emit stray RF electromagnetic energy that could be hazardous to health.

Table 4.3: Industries using dielectric RF heaters

Automotive	Plastic heat-sealing
Furniture and wood	Rubber products
Glass fibres	Textiles
Paper products	

Surveys of RF heaters indicate that they may be grouped into five categories: sewing-machine, shuttle tray, turntable, pressure-sealed applicator and edge-glue dryer. A summary (53) of operator exposures received from the various types of RF heater, time-averaged over the duty cycle (fraction of on-time to total time during normal operation), is given in table 4.4. Operators may be exposed to equivalent plane-wave power densities exceeding 10 mW/cm^2, especially from the sewing-machine and shuttle-tray type heaters.

RF heaters represent a major concern with respect to operator exposure. Values of electric and magnetic field strengths, even when averaged over the duty cycle, can still exceed all national exposure limits. Of particular concern is that RF heaters operate at frequencies in or near the region of human resonant absorption (60-100 MHz). Hence workers near RF heaters can absorb considerable amounts of stray energy. Contact (electrical) of the workers with a ground plane can shift the frequency at which maximum absorption occurs to well within the frequency band of most RF heaters. Account must be taken of obvious problems with these devices, such as RF burns and electric shock. Stray fields from the RF heaters can be relatively easily reduced by proper design and appropriate location of shields for the RF fields (49, 8).

4.2.6 RF induction heaters

These heaters are used in the production of skis and radio tubes, in metal processing, and so on, and operate in the low RF frequency range. Little or no survey data is available on these devices, although many of the same problems found with RF dielectric heaters could be anticipated.

Table 4.4: Exposures (corrected for average duty cycle) of operators of various dielectric (RF) heaters

Heater type	Number of units	Above 1 mW/cm^2		Above 10 mW/cm^2	
		-	%	-	%
Sewing machine	33	19	58	13	39
Shuttle tray	16	10	63	1	6
Turntable	4	2	50	0	0
Pressure-sealed applicator	21	0	0	0	0
Edge-glue dryer	7	0	0	0	0
All	81	31	38	14	17

Source: Stuchly et al. (53).

4.2.7 Medical applications (diathermy)

Diathermy is the application of RF or MW energy to generate heat in body tissues. Therapeutic diathermy is used in the treatment of many conditions, including pain, arthritis or muscle spasm, and in physical rehabilitation (restoration of tendon function, for example), by producing deep heating without excessive temperature elevation of the skin surface or in subcutaneous fatty tissue. Short-wave diathermy devices operate at frequencies from 10 to 100 MHz (predominantly at 27.12 MHz), while microwave diathermy equipment operates from 300 to 100,000 MHz (normally at 2,450 MHz, but sometimes at 915 MHz).

Therapeutic diathermy involves the use of one or two applicators, coils or conducting cables applying energy to the treatment area for a specified period of time. Measurements of leakage radiation from the applicators, made at 5 cm from the interface between a tissue-equivalent phantom and a microwave diathermy applicator indicated levels up to 44 mW/cm^2 (6). Exposures to the operators (standing at the control console) during microwave diathermy treatment could be up to 1.3 mW/cm^2 (50).

Leakage-radiation levels from short-wave diathermy devices represent a greater problem. Operators of these devices can be

Table 4.5: Operator exposures to leakage radiation during various short-wave diathermy treatments

Treatment number	Number of devices	Eye		Waist	
		E (V/m)	H (A/m)	E (V/m)	H (A/m)
1	2	4-16	0.05-0.09	6-63	0.08-0.12
2	2	6-22	0.07-0.09	5-55	0.08-0.15
3	1	315	0.3	315	0.5
4	1	24	0.09	25	0.23
5	3	2-40	0.08-0.12	2-70	0.09-0.4
6	5	2-7	0.05-0.07	2-8	0.05-0.09
7	6	2-22	0.05-0.08	2-84	0.05-0.23
8	3	3-12	0.05-0.08	6-100	0.09-0.23
9	3	16-71	0.14-0.35	25-123	0.5-0.95
10	4	2-20	0.05-0.1	3-55	0.08-0.4
11	1	32	-	40	-

Source: Stuchly et al. (53).

exposed to field strengths up to 0.47 A/m and 250 V/m (equivalent to free-space power densities of 8.3 and 16.6 mW/cm^2 respectively) (50). Results of operator exposures are shown in table 4.5 (53). It is evident that greater protection of the operator is needed. This can be in the form of shielding the RF cables (between the RF generator and the applicators) and educating the operator with regard to where and where not to stand during treatments.

Hyperthermia equipment, increasingly used in cancer treatment, is normally designed to operate at 915 MHz and 2,450 MHz, but may operate at other frequencies, e.g. 434 MHz. Screened rooms may be needed to limit radiation emissions beyond the patient treatment area.

4.2.8 Video display terminals

Video display terminals (VDTs) have found widespread application in industry, business and government offices, the communications field and as education aids. Their proliferation has given rise to questions about adverse health effects associated with their use, in particular about the possibility of VDTs posing a radiation hazard to operators of these devices.

A VDT basically consists of a television-type monitor screen, a keyboard and an electronic data-handling system. The VDT can display only information received through its data-handling system from a computer or word processor.

In common with many electronic devices, including television receivers, some VDTs can generate detectable levels of low-frequency electromagnetic radiation. This radiation is generally at frequencies below 10 MHz and is commonly referred to as an electrical "noise". The RF "noise" originates in the RF oscillator circuit around the "flyback" transformer, which the VDT must have to generate its visual display.

Where RF radiation from VDTs has been measured, the maximum levels are found close to the surface of the VDT case, in the vicinity of the RF oscillator circuit. The levels of RF were found to drop off very rapidly with distance from the surface, and at 20-30 cm away the RF was extremely low (5-10 V/m) (7, 42). The design of RF oscillators used in VDTs and the nature of their operation precludes any significant emission of microwave radiation.

4.3 Occupationally related exposure type and effects

This section deals with the man-made sources of radiation, which include any appliance that generates electricity or is driven by means of an electric current. Such appliances generate electromagnetic fields which propagate through space in the form of electromagnetic waves. Propagation of electromagnetic waves may be classified as unintentional, as in the case of fluorescent light tubes, and deliberate emitters, as in the case of radio broadcasting, radar installations and other electronic wireless communication systems.

4.3.1 RF bands exposure and effects

RF bands include the following:

(a) low frequency - LF (30-300 kHZ);

(b) medium frequency - MF (0.3-3 MHz);

(c) high frequency - HF (3-30 MHz);

(d) very high frequency - VHF (30-300 MHz).

A summary of the characteristics, uses and sources of exposure, category of exposed workers' population and health effects is presented in table 4.6.

Table 4.6: RF bands exposure and health effects

		Low frequency LF (30-300 K Hz)	Medium frequency MF (0.3-3 M Hz)	High frequency HF (3-30 M Hz)	Very high frequency VHF (30-300 M Hz)
Wavelength		10-1 km	1-0.1 km	100-10 m	10-1 m
Metric subdivision		Kilometric waves	Hectometric waves	Decametric waves	Metric waves
Photon energy (ev)		$10^{-10} - 10^{-9}$	$10^{-9} - 10^{-8}$	$10^{-8} - 10^{-7}$	$10^{-7} - 10^{-6}$
Uses and sources of exposure		Broadcasting, television, air traffic control, radar, radio navigation, power line carrier, marine and aeronautical mobile, medicine.	Communications, radio navigation, marine, radiophone, amateur radio, plastic pre-heating, microwave ovens, food industry, TV, radar, medicine.	Citizens' band, amateur radio, medical diathermy, AM broadcast, international communications, plastic pre-heating, altimeters, air- and ship-borne radar, navigation, satellite communication.	Police, fire, amateur radio, FM, VHF-TV, industrial RF equipment, FM broadcast, radio-meteorology, space research, nuclear physics, radio spectroscopy.
Category of exposed workers		Broadcasting, transmitter and television personnel, medical personnel, factory workers related to the above occupations.	Microwave testers, diathermy and microwave operators and maintenance workers, broadcasting transmitter and television personnel, electronic engineers and technicians, air crew, missile launchers, radar operators and technicians, medical personnel.	Physicists, microwave development workers, radar operators, marine and coastguard personnel, sailors, fishermen and persons working on board ships, physiotherapists.	Physicists, microwave development workers, radar operators, meteorologists.
Health effects observed in	Man	Central nervous system (CNS) drug tolerance. CNS functional human disorders. Functional changes in the nervous and cardiovascular system. Spermatogenesis, altered sex ratio of births, menstrual pattern changes, congenital effects in new-born babies, decreased lactation.			
	Animals[A]			Developmental abnormalities in rats	Hyperthermia Disturbance in bioelectric function of chick in vivo

[A] For details on the duration of exposure, power densities and other experimental conditions, reference should be made to the literature as space does not permit their inclusion here.

Table 4.7: MW bands exposure and health effects

		Ultra-high frequency UHF (0.3-3 G Hz)	Super-high frequency SHF (3-30 G Hz)	Extra-high frequency EHF (30-300 G Hz)
Wavelength		1-0.1 m	10-1 cm	10-1 mm
Metric subdivision		Decimetric waves	Centimetric waves	Millimetric waves
Photon energy (ev)		$10^{-6} - 10^{-5}$	$10^{-5} - 10^{-4}$	$10^{-4} - 10^{-3}$
Uses and sources of exposure		Short-range broadcast, communications, amateur radio, taxi, police, fire, radar, citizens' band, radio-navigation, UHF-TV, microwave ovens, medical diathermy, industrial heating, telemetry, troposcatter and meteorological radar.	Satellite communications, radar, amateur radio, microwave relay, airborne weather radar, shipborne navigational radar, microwave point-to-point.	Terrain avoidance, radar, airborne weather (cloud detection) radar.
Category of exposed workers		Broadcasting and communication personnel, operators of the above, medical personnel, physiotherapists.	Physicists, microwave operators, persons working on board ships.	Personnel engaged in the above.
Health effects observed in	Man	- CNS drug tolerance - CNS functional human disorders - Functional changes in cardio-vascular system - congenital malformations		
	Animals[A]	Developmental effects in rats Neural, blood-forming and immunological effects Convulsion in monkeys Hyperthermia in dogs Cataract in white rabbits Changes in neuroendocrine functions	Blood-forming and immunological effects Neural effects in rabbits Behavioural modifications Hyperthermia in rats Endocrine gland, hormone-level and thyroid function effects in rats	

[A] For details on the duration of exposure, power densities and other experimental conditions, reference should be made to the literature as space does not permit their inclusion here.

4.3.2 MW bands exposure and effects

MW bands include the following:

(a) ultra-high frequency - UHF (0.3-3 GHz);

(b) super-high frequency - SHF (3-30 GHz);

(c) extra-high frequency - EHF (30-300 GHz).

A summary of the characteristics, uses and sources of exposure, category of exposed workers' population and health effects is presented in table 4.7.

5. Community effects

As indicated in tables 4.6 and 4.7, there is a complex array of man-made radiation-emitting equipment with a varying degree of power densities and frequency ranges. Thus, depending on the mode of use, a large segment of the population receives exposure of varying degrees. The extent of this exposure is also dependent on obvious differences such as the development of the country and the degree of urbanisation. Exposure of the urban population to radiation in highly industrialised countries is in the order of W/cm^2 according to investigations by the United States Environmental Protection Agency. The levels of exposure among inhabitants of high buildings in the vicinity of the roof-top antennae of broadcasting and television stations may be in the order of a few hundred W/cm^2 to a few mW/cm^2 (63). Fifty per cent of the urban population of the United States is exposed to less than 0.005 mW/cm^2, 95 per cent to less than 0.01 mW/cm^2 and 99 per cent to less than 0.1 mW/cm^2 (57).

In summary, general population exposure may be regarded as involving long-term, very low-level, intermittent exposure to a very wide range of RF and MW radiation frequencies.

6. Instrumentation and measurement techniques

Instruments for the measurement of radio-frequency and microwave fields can be divided into two basic components, a sensor or probe and an indicator or meter. The sensor consists of an antenna in combination with a detector, and provides the signal which is carried to the meter. To accomplish this without causing perturbation, the leads may be high-resistance wires, or a fibre-optic link. The instruments may take the general form of signal conditioning circuitry, with a display device (meter). The performance and application of the instruments are largely determined by the sensor system.

Instruments in current use do not normally measure power density (mW/cm^2) as such, but calculate this from a measurement of the magnetic field (A/m) or the electric field (V/m). The magnetic field sensor is usually one or more loop antennae. Electric field strength sensors most commonly use one or more dipoles. A single-dipole sensor gives information about the polarisation of the field being measured, but the antenna must be rotated in three mutually orthogonal directions to determine the true field strength. With three dipoles in an orthogonal array, the response is isotropic (non-directional and non-polarised). Other less common configurations of antenna are spirals which are circularly polarised and therefore equivalent to two orthogonal dipoles, and thin resistance films deposited on the surface of a sphere (41).

6.1 Measuring instruments (thermocouples, diodes, bolometers, fluorescent tubes)

The two types of detectors in most common usage are thermocouples and diodes. These respond to the square of the voltage induced in the antenna and are scaled linearly, in terms of power density. Diode detectors can use a smaller sensor but may have a limited square response region (which may require biasing) and may become inaccurate at high field strengths. This problem has been overcome in some instruments by the use of a microprocessor. Thermocouple and bolometric devices both use heating by RF and MW to measure accurately, even at very high instantaneous field strengths. Bolometric instruments tend to be very robust, as they are inherently self-protecting against overload (high fields outside the instrument's measurement capability). Although thermocouple instruments can have the widest frequency response, the lowest sensitivity to environmental temperature and the lowest zero drift, they are normally very susceptible to overload or burn-out in high instantaneous fields. A number of commercially available instruments are capable of accuracies of ±1 dB or better within their useful ranges.

Small fluorescent tubes have been used in some cases to detect RF radiation around equipment. Although these tubes do light up when exposed to RF radiation, they are unable to provide any measure of the power density of radiation emitted.

6.2 Radiation measurement and survey instrument selection

Prior to surveying RF and MW fields it is important to determine as many of the known characteristics as possible. In many cases it may be appropriate to obtain some estimate of the intensity of the fields given off by various sources. Formulae for making these calculations can be found in references 3 and 9. This will permit a more appropriate selection of survey instruments and test procedures. A check list may include:

(a) generated power;

(b) carrier frequency (or frequencies);

(c) modulation characteristics - peak, average values, etc.;

(d) type of antenna and its gain, beam width, polarisation and scanning programme;

(e) distance from the source of site surveyed;

(f) existence of absorbing or scattering objects likely to influence the radiation measurements;

(g) spurious radiation frequencies that may be emitted by the source, including harmonics;

(h) other interfering sources.

A review of such check lists, though rather elementary, is a necessary discipline if inspectors are to avoid some simple but often surprising situations. There is the possibility that spurious electromagnetic fields may exist at the survey site and produce radio-frequency interference (RFI) in the instrument electronics unless the instrument is designed for operation in the presence of such spurious frequencies. If the information is well defined, inspectors may, after making estimates of expected field strengths and selecting the appropriate survey instrument, carefully proceed with the inspection. Obviously they should avoid entering RF fields that present the risk of being exposed to levels above the recommended limits. Care must also be taken with pulsed fields to avoid burn-out of the detector.

Generally speaking, estimates of the intensity of radiation are made under the following field conditions.

6.2.1 Far-field exposure

The far-field exposure refers to situations where such exposure is at a distance greater than about one wavelength from the source. In this case, the electric (E) and the magnetic (H) field strengths, expressed by V/m and A/m respectively, are said to be in phase and yield a product of VA/m^2 which may be expressed as W/m^2 (see section 2.2).

In the above case, the power density can be calculated from the formula:

$$Pd = GPt/4\pi d^2 = AeP/\lambda^2 d^2$$

where:

Pd = power density

G = the far-field gain (representing the ratio of the observed power density, on axis, to the power density from a point source having the same output power and emitting equally in all directions.

Pt = the power delivered to the antenna (W)

d = the distance from the antenna (m)

λ = the wavelength (m)

Ae = the effective area of the antenna (m^2).

6.2.2 Near-field exposure

The near-field exposure refers to situations where such exposure is at a distance less than that of the wavelength from the source. In this case, the reactive components of E and H become progressively more important. At this range power density calibrated instruments become more inaccurate.

For parabolic antennae, in the above case, the power density can be calculated from the formula:

Pd = 4Pt/Ae

where:

Pt = transmitted power.

Survey instruments used in occupational exposure determinations should be accurate and capable of measuring the lowest power density limits required by appropriate exposure or emission standards, with an accuracy as specified in any applicable standard or by a control authority, or at least 3 dB, or better (or the equivalent electric or magnetic field strength with an accuracy of 1.5 dB, or better). A survey instrument should be calibrated against a secondary standard instrument at least once a year, or after any repair that may affect its calibration.

6.3 Survey instrument characteristics

Prior to purchasing a survey instrument for RF surveys, the following features should be considered:

- the range of readings to be made;

- the response time of the meter (this should not normally exceed 3 s);

- the meter non-linearity (this should not exceed 20 per cent).

- the environmental temperature sensitivity (should not exceed 1 dB in the range 4-30°C);

- the detector response (should be isotropic with a maximum deviation not exceeding 1 dB);

- change in response due to supply voltage variations (should not be greater than 20 per cent);

- the susceptibility to interference by signals outside the frequency band being measured (should be minimal);

- the zero drift (should not exceed 10 per cent per hour); and

- the sensitivity to overload (should be at least a factor of 10), including the possibility of damage from high instantaneous power from pulsed fields of low pulse-repetition frequency.

The overall inaccuracy of the survey instrument, with all the above factors taken into account, should not exceed ±1 dB.

7. Occupational exposure limits

Occupational exposure limits aim at protecting healthy workers who are known to be subject to exposure to radiation and as such are under medical surveillance. They are different, for example, from general population standards in that setting of the latter takes into consideration a variety of factors such as the health and susceptibility of the individual, as well as interaction with the ambient atmosphere and the lengthy duration of exposure.

7.1 Basis for setting RF and MW radiation protection standards

Biophysical considerations, animal studies and limited human occupational exposure data constitute the basis for the establishment of health protection standards. In establishing health protection standards, different approaches and philosophies have been adopted in various countries.

Pursuant to the discussion on the dose-effect relationship (section 3.1.4), there are several variables to be considered when determining exposure safety factors. These are:

(a) the absorption of RF electromagnetic energy by humans of various sizes, with particular reference to whole or partial body resonant absorption of energy;

(b) the lack of knowledge of relating peak SAR (non-uniform energy absorption) to observed biological effects;

(c) environmental conditions - the exposure limits should be protective under adverse conditions of temperature, humidity and air movement;

(d) the reflection, focusing and scattering of the incident fields in such a way that enhanced energy absorption occurs;

(e) possible altered response of humans taking medicines and being exposed to RF fields;

(f) possible combined effects of RF electromagnetic energy with chemical or other physical agents in the environment;

(g) the confirmed effects of modulated microwave fields on the central nervous system and the possible existence of "power" and "frequency" windows for effects;

(h) the effects of RF radiation exposure on persons having decremental health conditions or illness, including immuno-regulatory problems, epilepsy, etc.;

(i) the effects of modulated (pulsed) RF fields on cellular function (cell-membrane interaction); and

(j) the effects of long-term, low-level exposures versus short-term high level exposures.

At present the data on (e), (f), (g), (h), (i) and (j) are insufficient to determine if these effects present a potential health concern.

For frequencies below 10 MHz, very little information is available on biological growth and wound healing. The rate of energy absorption by the human body rapidly decreases at lower frequencies. However, because one cannot necessarily assume that a biological effect is directly related to the amount of energy absorbed, added safety margins should be incorporated into the exposure limits than would be suggested from the whole-body-averaged SAR for humans. The ANSI Committee (59). for example, chose a whole-body, time-averaged upper limit of 0.4 W/kg, with a peak SAR not to exceed 8 W/kg when averaged over a time of 6 min for 1 g of tissue.

7.2 Exposure standards and recommended limits

Few countries have promulgated legal standards that limit RF exposure. Standards not having a legal backing are generally referred to as protection guides or safety guide-lines. An example of this is in the ANSI standard C95 on RF radiation exposure (59).

Exposure standards for the general population present a more complex problem than those for occupational exposure, since factors which can be controlled in occupational situations are not always amenable to similar control.

Standards for controlling exposure to RF radiation were introduced in the 1950s in the United States and the USSR. The USSR (61) standard was derived from experiments on small laboratory animals and surveys of people occupationally exposed. Functional changes were reported in animals after exposure to power densities of about 1 mW/cm^2 in the frequency range of 1 to 10 GHz for periods of greater than one hour. One-tenth of this value was recommended as the safe level for exposure throughout the working day, and an extra safety factor of 10 was applied to allow for the assumed individual variation in sensitivity. Taking into account the human studies and safety factors, this yielded an occupational exposure limit of 10 $\mu W/cm^2$ for the working day. The USSR standard allows incremental increases in exposure, each by a factor of 10, for exposure durations shorter than two hours, and for exposure durations shorter than 20 minutes, respectively (see table 7.1). A later amendment to this standard and its interpretation (Akimenko - personal communication) suggest that for continuous exposure the permissible level of incident power density is derived from the formula:

$$p = \frac{200}{t} \ \mu W/cm^2$$

where t is expressed in hours.

For intermittent exposure the following formula is used:

$$p = \frac{2,000}{t} \mu W/cm^2$$

Unofficial interpretation is that 200 $\mu W/cm^2$ exposure for 1 hour is permissible (averaged). The ceiling level is 1 mW/cm^2.

The United States (59) ANSI C95 voluntary standard was developed following a review of the current literature and is based on a determination of the SAR of electromagnetic energy by models of the human body over the frequency range of 300 kHz to 100 GHz. In essence, this literature review found that no substantiated short-term adverse health effects appeared to occur from exposure conditions producing an SAR of less than 4 W/kg. The final limit was derived by incorporating a safety factor of 10 to allow for possible long-term effects (see table 7.2).

Most of the biological investigations relevant to the development of exposure standards have been carried out over the frequency range of 1 GHz to 10 GHz. There is, however, a considerable divergence in the frequency range over which the standards of individual countries apply. Reference should generally be made to the standards themselves for detailed information but tables 7.1 and 7.2 provide summaries of some existing standards. Table 7.1 summarises the occupational exposure standards presently existing in the USSR (61), Czechoslovakia (15) and Poland (44, 45). Standards similar to those in the USSR exist in Bulgaria (17) and the German Democratic Republic (16). Table 7.2 summarises the occupational exposure standards now existing in the United States (59), Canada (12) and Sweden (54). Information on the applicable frequency range, exposure limits, exposure duration and RF source characteristics is provided for comparison purposes. Exposure standards for MW and RF radiation are presently being proposed in Australia (14), the Federal Republic of Germany (20) and the Netherlands (58).

7.3 International exposure limits

A review of national exposure standards for non-ionising radiation has been made by the WHO/UNEP/IRPA (63), Michaelson (39) and the International Electrotechnical Commission - IEC (28). The Commission of the European Communities (19) proposed a 10 mW/cm^2 standard for exposure to microwaves (300 MHz to 300 GHz). The International Radiation Protection Association's International Non-Ionising Radiation Committee (IRPA/INIRC) has drafted an international standard for NIR from 100 kHz to 300 GHz (34). This is the only truly international attempt to preparing a comprehensive occupational exposure standard. Details of this proposal are provided in table 7.3.

The IRPA recommends that occupational exposure to RF radiation at frequencies at or above 10 MHz should not exceed a SAR of 0.4 W/kg when averaged over the whole body, or 4 W/kg when averaged over any one gram of tissue. For RF radiation exposure at frequencies below 10 MHz, the levels of unperturbed root mean square (RMS) electric or magnetic field strength should not exceed the values given in table 7.3.

The limits of occupational exposure given in table 7.3 for the frequencies between 10 and 300,000 MHz are the working limits derived fom the SAR value of 0.4 W/kg. They represent a practical approximation of the incident plane-wave power density needed to produce the whole-body average specific absorption rate of 0.4 W/kg.

Table 7.1: USSR, Poland and Czechoslovakia: Occupational exposure standards

Standard	Frequency	Exposure limit	Exposure duration	CW/pulsed	Antenna stationary/ rotating	Remarks
USSR (ref. 61)	-30 MHz	20 V/m	working day	both	both	Military units and establishments of the Ministry of Defence excluded
	30-50 MHz	10 V/m	working day	both	both	
		0.3 A/m	working day	both	both	
	50-300 MHz	5 V/m	working day	both	both	
	0.3-300 GHz	10 U W/cm^2	working day	both	stationary	
		100 U W/cm^2	working day	both	rotating	
		100 U W/cm^2	2 hours	both	stationary	
		1 mW/cm^2	2 hours	both	rotating	
		1 mW/cm^2	20 min	both	stationary	
Czechoslovakia (ref. 15)	30 kHz-30 MHz	50 V/	working day	both	both	
	30-300 MHz	10 V/m	working day	both	both	
	0.3-300 GHz	25 W/cm^2	working day	CW	both	
		10 W/cm^2	working day	pulsed	both	L = power density in MW/cm^2
		L (W/cm^2) x	t (hours) = 200	CW	both	t = exposure duration in h/day
		L (W/cm^2) x	t (hours) = 80	pulsed	both	
Poland (ref. 44)	0.3-300 GHz	0.2 mW/cm^2	10 hours	both	stationary	
		0.2-10 mW/cm^2	$32/P^2$ (hours)	both	stationary	P = power density in W/m^2
		1 mW/cm^2	10 hours	both	rotating	
		1-10 mW/cm^2	$800/P^2$ (hours)	both	rotating	P = power density in W/m^2
Poland (ref. 45)	10-300 MHz	20 V/m	working day	both	both	
		20-300 V/m	$3200/E^2$ (hours)	both	both	E = electric field intensity

Table 7.2: United States, Canada and Sweden: Occupational exposure standards

Standard	Frequency	Exposure limit	Exposure duration	CW/pulsed	Antenna stationary/ rotation	Remarks
US (ref. 59)	300 kHz-3 MHz 3 MHz-30 MHz 30 MHz-300 MHz 300 MHz-1.5 GHz 1.5 GHz-100 GHz	100 mW/cm^2 $900/f^2$ mW/cm^2 1 mW/cm^2 f/300 mW/cm^2 5 mW/cm^2	no limit	both	both	f = frequency in MHz (averaged over any 0.1 h period); equivalent power density given; standard expressed in $(V/m)^2$ and $(A/m)^2$
Canada (ref. 12)	10 MHz-1 GHz	1 mW/cm^2 60 V/m 0.16 A/m	no limit	both	both	averaged over 0.1 h period
	1-300 GHz	5 mW/cm^2	no limit	both	both	
	10 MHz-300 GHz	1-25 mW/cm^2	60/X (min)	both	both	X = power density in mW/cm^2
	10 MHz-300 GHz	25 mW/cm^2 (maximum)	2.4 min	both	both	
Sweden (ref. 54)	0.3-300 GHz	1 mW/cm^2	8 hours	both	both	averaged over 0.1 h
	10-300 MHz	5 mW/cm^2	8 hours	both	both	averaged over 0.1 h
	10 MHz-300 GHz	25 mW/cm^2 (maximum)	8 hours	both	both	ceiling exposure level averaged over 1 s

Table 7.3: Limits for whole body occupational exposure (Recommended by the International Radiation Protection Association)

Frequency range (MHz)	Unperturbed RMS electric field strength (V/m)	Unperturbed RMS magnetic field strength (A/m)	Equivalent plane-wave power density	
			W/m^2	mW/cm^2
0.1-1	194	0.51	100[a]	10[a]
1-10	$194/f^{1/2}$	$0.51/f^{1/2}$	100/f[a]	10/f[a]
10-400	61	0.16	10	1
400-2 000	$3\ f^{1/2}$	$0.008\ f^{1/2}$	f/40	f/400
2 000-300 000	137	0.36	50	5

Notes: (1) f = frequency in MHz. (2) The limits in the frequency ranges above 10 MHz may be exceeded for specific applications provided the SAR remains below 0.4W/kg when averaged over the whole body, and the spatial average SAR remains below 4W/kg when averaged over any one gram of tissue. The limits for frequencies at or below 10 MHz may be exceeded (up to 630 V/m or 1.6 A/m) provided workers take the necessary precautions to prevent potentially severe RF burns.

[a] These values are provided for information only and are not to be considered for determining compliance unless measurements are made in the far field.

Source: IRPA (34).

The limits apply to whole-body exposure from either continuous or modulated electromagnetic fields from one or more sources, averaged over any 6-minute period during the working day (8 h per 24 h).

Although very little information is presently available on the relation of biological effects with peak values of pulsed fields it is suggested that the instantaneous peak values for all frequencies should not exceed 100 times the limits in table 7.3 for the frequency concerned and, further, that the limits of occupational exposure, averaged over any 6-minute period, should not be exceeded.

7.4 Source related (emission) standards

Emission standards or device standards which, among other safety requirements, limit the emission of leakage radiation from a particular class of device, should normally be derived from exposure standards. The latter represent the "safe" level of human exposure to RF radiation, and emission standards should then require lower radiation-leakage levels to account for multiple sources of exposure. It is a fundamental responsibility of each country to develop appropriate emission standards for hazardous RF-emitting devices.

The IEC (31) has published a standard requiring exposure from radio-transmitting equipment to be limited to 10 mW/cm^2 over the frequency range of 30 MHz to 30 GHz. However, the best known emission standards concern the limits on microwave radiation leakage from microwave ovens. Limits of emission for microwave ovens have been published by the IEC (29) and are similar in most countries.

Emission standards should take into account the possibility of human access, both routine and for servicing equipment. By means of positive control of human access, substantial savings in equipment layout may be accomplished.

8. Control of and protection from exposure to RF and MW radiation

This section recommends measures for the effective control of and protection from exposure to RF and MW radiation.

8.1 Role of the competent authorities

The competent or regulatory authorities whose terms of reference and functions are concerned with protection against the harmful effects of RF and MW radiation should be required to co-operate with one another. This co-operation is necessary to ensure that each authority, or department within an authority, is well aware of the other organs' responsibilities so as to avoid duplication of effort. Such a role is envisaged in the following manner.

8.1.1 Setting of the relevant rules, regulations, standards and codes

Given that effects have been reported at power densities (irradiance exposure) too low to be calorimetrically significant, and that occupationally exposed workers should be subjected to occupational surveillance, both physical and medical, the competent authorities have the urgent responsibility to formulate the necessary legislation and/or regulations for radiation protection. This should be undertaken in consultation with the representative organisations of the employers and workers concerned. In addition, the authorities should provide, as necessary, detailed specific guidance for design, manufacture and use of the various radiation sources. Such regulations should also specify the relevant rules, standards and occupational exposure conditions.

8.1.2 Licensing, notification and/or registration systems

Because more data is needed to clarify the interaction mechanism and to determine the threshold levels of effects, the competent authorities should consider the application of the principle that exposure of workers (and the public) to MW and RF radiation should be kept as low as reasonably possible, and that the exposure standards should never be exceeded. The classification of sources of exposure or practices according to notification, registration, or licensing requirements should be based on the health hazards involved. Whenever more rigid control is necessary, then the authorities should specify the exposure sources requiring a licence, together with the procedures and conditions for obtaining one. The licence permit should include follow-up methods to ensure compliance during plant construction, commissioning and operation, including future modification of design or working procedures. It should be remembered that the granting of a licence should not preclude a change during the period of its validity. A request for such a change may be initiated by the licensee, or the change may be desirable or necessary as a result of experience gained during operation at the workplace or elsewhere, or as a consequence of technological innovation or of safety research and development. The licensing process should define the various responsibilities concerned with the planning, design, construction, commissioning and operation of the facility. The licensee on the other hand should submit and make available to the competent authorities all the information requested. Whenever a significant change in operation or use of the exposure source occurs relevant to the licensing process, the licensee should submit that information, preferably in accordance to a specified procedure, to the authorities.

8.1.3 Quality control requirements for the design, planning and construction of plant and equipment

Quality control programmes for each constituent area of activity in the design, planning and construction of plant and equipment used depends on its magnitude and the hazards involved and may vary from a very extensive, multi-phase and multi-step plan to a rather more simple and straightforward operational procedure. It should be recognised that the basic responsibility for achieving quality in performing a particular task (e.g. in design, manufacturing, commissioning or operation) rests with those assigned to it and not with those seeking to verify that it has been carried out correctly.

Whenever necessary, the competent authorities should establish the health standards to be applied to the manufacturing and supply of materials, machines, tools, equipment, other implements and installations, including the construction of the latter.

Finally, the authorities should encourage manufacturers to undertake studies and research in order to improve the design, construction and performance of the equipment, materials and facilities supplied so as to contribute to the minimisation of occupational hazards.

8.1.4 Inspection and intervention

The competent authorities should establish a system of inspection to supervise safety precautions to ensure compliance with the relevant standards and with the requirements as specified in the licence. They should also assume the necessary power to intervene in cases of non-compliance with the standards. Any situation which has resulted or is expected to result in doses in excess of the exposure limits should be reported in an approved manner.

8.2 Responsibility of the employer

The owner of the device or installation emitting RF radiation:

(a) is ultimately responsible for the radiation safety of employees;

(b) is responsible for purchasing or providing RF equipment which meets all appropriate governmental standards when new and during its lifetime of use;

(c) is responsible for ensuring that the RF equipment meets appropriate radiation safety standards and safety requirements as specified in this document;

(d) may delegate this responsibility depending on the size of the organisation and the amount of radiation-emitting equipment used. One or more persons should be designated by the owner to carry out the role of responsible user and RF safety inspector;

(e) is responsible for action to reduce the exposure of workers to RF radiation and for the organisational arrangements required to prevent the risks associated with exposure;

(f) should establish and publicise (preferably in writing) a general policy emphasising the importance of prevention, and should take the decisions and the practical steps required to give effect to national regulations and to this document.

8.3 Responsibility of the radiation protection officer

There should be at least one person, or a safety committee, designated by the owner or employer as the responsible user, to undertake responsibility for:

(a) establishing safe operating procedures for the equipment;

(b) ensuring that all radiation workers are aware of the contents of this document and knowhow to put it into practice;

(c) setting radiation safety rules and ensuring that all staff are made aware of them;

(d) ensuring that assigned equipment is maintained and used correctly by competent personnel;

(e) knowing the exposure levels in the vicinity of the equipment under normal conditions of use;

(f) defining areas where exposures in excess of the recommended limits may result, and posting warning signs that clearly indicate the permitted occupancy durations;

(g) ensuring that appropriate radiation surveys or monitoring are performed when and as required, and maintaining records of such monitoring;

(h) investigating and reporting exposures which may be in excess of the limits recommended;

(i) designating staff as RF workers;

(j) arranging for the medical examination and treatment, if necessary, of "over-exposed workers" as deemed appropriate;

(k) recording levels and duration of exposure for persons who have been exposed in excess of the recommended limits.

8.4 Responsibility of the health physicist

A health physicist may be delegated some of the responsibilities of the responsible user. In small organisations, the responsible user and the health physicist may be the same person. This person should have direct access to the owner or employer of the organisation. The extent and nature of the responsibilities will depend on the size of the organisation and the number of devices. In general, the health physicist provides technical support in the planning of the installation and the operation of RF devices.

Health physicists should have the following responsibilities:

(a) if called upon to supervise compliance with the regulations, to take into consideration the provision of this document;

(b) to ensure that an effective safety programme is developed and instituted whenever there is a special hazard due to RF radiation exposure;

(c) to attach special importance to proper briefing of workers, and to co-operation between employer and workers in the reduction or prevention of RF radiation exposure;

(d) to ensure that the responsible user or safety committee receives the information needed to be effective;

8.5 <u>Responsibility of the worker</u>

RF workers are responsible for the day-to-day operation and/or maintenance of RF emitting devices and must:

(a) understand the hazards associated with operating the specific RF devices assigned to them, and in particular, the importance of:

  (i) any interlock systems and dangers associated with defeating such systems; and

  (ii) adherence to all occupancy restrictions;

(b) be able to recognise malfunctions of the specific devices assigned to them that may result in high radiation exposures;

(c) be aware and trained in normal safe operating practices and the procedures to be followed in the event of malfunction of the devices, or in an emergency situation arising from excessive radiation leakage;

(d) be willing to undergo any reasonable prescribed medical surveillance; and

(e) use protective equipment provided, as necessary.

8.6 <u>Responsibility of vendors and suppliers</u>

Equipment manufacturers are responsible for making equipment that conforms to the appropriate standards within the country and for providing information on the hazards of operating and servicing RF equipment, sufficient to alert the owner or employer of the magnitude of the risk and the appropriate precautions that need to be taken.

8.7 Co-operation

The following principles are recommended with regard to co-operation between employers, workers, safety committees, manufacturers and so on:

(a) the employer should secure the workers' co-operation in action to protect their health and to reduce RF exposure, and should establish, by joint agreement, instructions and recommendations for the prevention of such exposure;

(b) employers and workers should co-operate in devising and implementing programmes for the prevention and control of RF exposure, particularly in conjunction with any safety committees to monitor the working environment;

(c) co-operation should be established between manufacturers and purchasers of equipment with a view to reducing unintended RF emissions of such equipment; and

(d) co-operation in the development of both voluntary and regulatory equipment standards should be considered.

8.8 Surveillance and monitoring of the workplace

The objective of radiation surveillance is to determine from records whether the equipment or installation complies with recommended standards of performance and personnel exposure (i.e. whether or not excessive exposure has occurred), to delineate boundary areas requiring shielding and to identify controlled and uncontrolled areas in the workplace (before the source(s) is operational).

8.8.1 Survey procedures and data collection

When conducting radiation surveys, the following procedure is recommended:

(a) radiation surveys should be carried out by a competent person, preferably the health physicist;

(b) surveys should be conducted in the following situations:

   (i) before routine operation begins, for all new installations capable of producing RF radiation exceeding the limits recommended;

   (ii) following any repairs or changes in working conditions, protective shielding and barriers that may affect the exposure levels, to ensure that the levels do not exceed the limits recommended;

(iii) when any malfunction is suspected that may affect radiation levels;

(iv) at two-year intervals at installations capable of exposing personnel in excess of the limits recommended;

(c) survey instruments must be selected to match the operating conditions and should be calibrated at least once every year. Their calibration should also be checked against another instrument before carrying out a radiation survey;

(d) during a radiation survey, a complete record should be kept of the field parameters (electric field strength, magnetic field strength, power density) at each worksite in order to make a realistic evaluation of potential personnel exposure;

(e) the direction of radiation propagated by the source(s) should be ascertained, because the body of the health physicist must not come between the source and the detector;

(f) during the inspection of any RF emitting device or installation, all safety interlocks, "ON-OFF" control switches, and required warning signs, labels and tags should be examined.

Records should be kept of all formal radiation survey measurements and their evaluation, including: the date that measurements were made, the number and types of devices in the area surveyed, the geometrical co-ordinates of the site surveyed, the radiation levels measured, and the model and serial number of the radiation survey instrument(s) used (see Appendix 3). Such records should also include a review of all known radiation incidents and their attributed causes.

8.8.2 Assessment and interpretation of occupational exposure data

In assessing and interpreting the occupational exposure data, the following steps are necessary:

(a) the results of occupationally monitored workers should be evaluated in terms of the available exposure limit (Chapter 7);

(b) for workers, whole or partial body exposure to continuous or pulsed MW or RF radiation having average power density within the range 0.1-1 mW/cm^2 include a high enough safety factor to allow continuous exposure from any part of the frequency range over the whole working day;

(c) higher exposure than the range 0.1-1 mW/cm^2 may be permissible over part of the frequency range for intermittent or occasional exposure;

(d) special considerations may be required in the case of pregnant women. The possible greater susceptibility of the embryo or foetus to MW or RF radiation must be taken into account;

(e) workers who were subject to exposures significantly higher than the recommended limits should be referred to the occupational physician for special medical surveillance. The exposure level alone does not necessitate the removal of the worker from work.

8.8.3 Implementation of recommendations

Exposures above the limits are unjustified and must therefore be prohibited.

Because of the lack of precise data on the non-thermal effects and their biological significance, it is recommended that:

(a) the exposure limits should be considered as the constraints for radiation protection of the worker, i.e. exposures at or below the limits are not automatically accepted as absolutely safe for purposes of planning and design and operational procedures;

(b) exposure should be kept as low as reasonably achievable (ALARA) below the limits. In such an approach economic and social factors are to be taken into consideration. A method of implementing ALARA would be cost-benefit analysis. Since it is complex to implement such an analysis in a quantitative form, it is recommended that a qualitative approach is considered.

8.9 Control of occupational exposure

The responsibility for the protection of workers and all other persons against the potentially adverse effects of exposure to RF fields should be clearly assigned to a department, agency, committee or individual, as indicated in the previous sections of this chapter. Such responsibility should include:

(a) the development and adoption of exposure limits and area control procedures and the implementation of a compliance programme. Equipment performance and emission standards limiting exposure from individual devices should then be derived from the exposure limits;

(b) the evaluation of protection against emissions from existing or newly manufactured equipment and installations. The development of an emission standard for the equipment may be considered, so as to avoid the dependence for safety on instructions and labels alone. Safe design of equipment should be required so that the need for use of additional protective devices is reduced to an unavoidable minimum;

(c) the development of standardised measurement procedures and survey techniques. Survey reports should be retained and should include details of the exposure conditions and, in cases where exposure limits are exceeded, indications of ways and means for reducing levels.

8.9.1 Controlled and uncontrolled areas

Areas where surveys indicate that exposure limits can be exceeded should be designated as areas of controlled access (see section 8.9.8) requiring the development of safety procedures to ensure that no one can be exposed to more than is allowed by the appropriate exposure standard. An example of where warning signs should be located is in the vicinity of high power communications systems. These are often unmanned, so it is necessary to provide fencing to prevent unauthorised entry to places where over-exposure could occur. Access should be restricted to workers who understand the possible consequences of over-exposure, who have a need to enter the area and for whom adequate supervisory controls are provided.

There are different means of controlling RF exposure in an area, e.g. by keeping the exposure levels down or remaining in the area for as short a time as possible to complete the task. Thus controlled areas should have warning signs that indicate the exposure level and the maximum time that should be spent in the area.

Uncontrolled areas are those areas that surveys have indicated to be such that exposure limits cannot be exceeded.

8.9.2 Engineering control

Exposure is preferably controlled by engineering measures alone, such as total or partial enclosure of the microwave fields, with leakage from the enclosure limited to that permitted by device emission standards. Where these measures are not sufficient to limit exposure to acceptable levels, additional shielding or screening materials can be provided (for more information on shielding effectiveness of screening materials see references 12 and 52) devices and installations sited in such a way as to keep exposure to personnel at levels as low as reasonably practicable, or restrictions placed on the occupancy of high-level exposure areas.

An effective way to control some occupational exposure situations is to either turn off or at least substantially reduce the power while maintenance work is carried out.

8.9.3 Supervisory control

The implementation of a complete radiation safety programme is a necessary first step towards reducing and controlling exposure. This includes designing a set of safe operating procedures for each item of RF equipment, controlling access to high exposure areas, setting up a schedule of radiation surveys and maintaining records.

8.9.4 Protective clothing

Basic protective measures include the provision of shielding or absorbing enclosures for appropriate components of generators. Personal protective equipment may also be used (e.g. gonad shields, protective suits, wire-netting helmets). Special protective goggles (metallised glass or wire-netting) are available, although they may provide insufficient protection.

Strong RF fields may induce dangerous voltages in metallic structures or electrical circuits quite unrelated to the application involved, especially when these structures or circuits unintentionally approach resonance at the frequency used (e.g. building frames, detonator circuits in quarries, wire fences, etc.). These voltages may in turn give rise to sparks and attendant hazards, particularly in connection with the use of explosives or flammable substances. If proper shielding is not possible, the radiating generators must be switched off during such use.

In general, modern protective measures, when properly applied, should ensure the safe operation of RF radiation sources (32).

8.9.5 Siting and installation

To assess the protective measures needed for locations in which RF and MW devices are routinely used, it is necessary to determine the occupancy and ownership of the surroundings. When conducting surveys around these devices, it should be noted that:

(a) radiation exposure in uncontrolled areas should not be allowed to exceed the limits recommended for such areas,

(b) radiation levels should be known and periodically checked as they may change in controlled areas where restricted occupancy is allowed. These areas should be designated according to section 8.9.8 and the maximum occupancy time posted;

(c) the immediate vicinity of unmanned, high-power sources of RF and MW radiation (e.g. some radio or television transmitting systems) should be fenced off and warning signs posted; it should be guarded (if necessary) to prevent unauthorised access to places where over-exposure could occur;

(d) RF and MW devices should be positioned as far away as is necessary from areas normally occupied by non-radiation workers;

(e) the siting of a RF or MW device should take into account the possibility of multiple exposures from RF radiation fields and leakage from other RF radiation sources in the vicinity;

(f) there should be no unnecessary metallic objects near any radiating RF or MW device. The presence of such objects may result in high intensity fields in some locations, and increase the possibility of RF shocks and burns;

(g) shielding or screening should be achieved by use of absorptive rather than reflective materials. When screening sources with enclosures not designed for occupancy, reflective materials such as screening or sheet metal may be used. It should be noted that the effectiveness of screening or attenuating materials is dependent on the frequency of the incident radiation. Attenuation properties of some common construction materials are given in references 52 and 12;

(h) all RF and MW devices should comply with applicable standards of design construction and performance specified in appropriate regulations;

(i) some high voltage RF generating devices may emit X-rays. Protection against X-rays should be accounted for (see Appendix 4).

8.9.6 Confined sources

Totally enclosed or confined beam sources such as microwave ovens and microwave blood warmers need special attention to ensure that:

(a) they comply with appropriate standards;

(b) adjustment of applied voltages, replacement of the MW power-generating components, dismantling of components and refitting of waveguides are undertaken only by properly trained personnel;

(c) special care is taken to ensure that no damage occurs to the part of the device making contact with the door or door seals;

(d) the main power is disconnected before reaching into any accessible openings;

(e) the door interlocks are fail safe;

(f) no microwave power-generating component is tested without an appropriate load connected to its output. The power generated should never be allowed to radiate freely into occupied areas; and

(g) higher power devices incorporate a key-switch on the power to the device to prevent unauthorised use.

8.9.7 Unconfined sources

Unconfined sources or open beam devices are devices that do not contain their own enclosure to shield the RF radiation. The following recommendations apply:

(a) an unconfined source should be operated only at a site or in a room which has been specifically designed to accommodate it;

(b) radiated energy should not be directed into occupied areas. Where this is not feasible, appropriate precautions must be followed to ensure that the exposure of personnel does not exceed the limits recommended. RF radiation will be reflected at metal surfaces and become secondary sources. Further, ungrounded conductors can become charged in an RF field since the metal structure acts as an antenna. At low RF frequencies, these charged structures can become sources of RF shocks and burns. In this regard it should be noted that calculations of the electrical currents induced in the human body making contact with ungrounded metallic objects suggest that RF shocks and burns may result from exposure to fields of approximately 615 V/m(100 mW/cm^2) in the frequency band 0.3 to 3.0 MHz (24). In a MHz field, humans touching an ungrounded metal object such as a car will (by induced currents) perceive an E-field of unperturbed strength above 80 V/m and let-go of the object (such as a compact car) above 200 V/m. For larger ungrounded metal objects the let-go field strength decreases. The unperturbed E-field strength causing a let-go situation remains approximately constant for a given object, for frequencies between 100 kHz and 10 MHz (24). Although occupational exposure to 200 V/m at frequencies below 1 MHz can result in minor RF shocks, workers could be trained to avoid these. Under special circumstances higher exposures (up to 600 V/m) could be tolerated by workers provided appropriate precautions were taken to avoid potentially severe RF burns. As a general rule, however, it is felt to be an unreasonable strain on workers to have to work continuously in such environments;

(c) where the power density produced by an open beam device exceeds ten times the appropriate exposure limit in locations accessible to people, the door to the room where the device is located should be interlocked in such a way that the device cannot be switched on when the door is open. If the door is opened when the device is on, the interlock should cause the microwave power to be turned off. It should always be possible to open the door from inside the room, and occupants should be positively warned to escape if the source is turned on;

(d) for any installation which is in operation for more than one month, interlock systems and all barriers controlling areas of restricted occupancy should be permanently installed to prevent accidental access to the zone of denied occupancy (see section 8.9.8).

8.9.8 Warning signs and labelling

The three warning signs shown in figure 8.1 are examples (12) of signs that could be used to indicate the nature and degree of hazard associated with a given device or location. Similar safety signs have been developed by the Commission of the European Communities (13) and France (22), but not specifically for non-ionising radiations. The nature of the radiation hazard is indicated by the symbol within the sign, and the degree of hazard is indicated by the shape and colour of its border. The size of the sign should be such that it is clearly distinguishable under the appropriate working conditions, being either illuminated or employing reflective materials as necessary.

It is suggested (12) that the CAUTION sign be BLACK on a contrasting colour such as a YELLOW background, the WARNING sign BLACK on, for example, an ORANGE background, and the DANGER sign RED on a WHITE background. Where personnel are employed at night or in darkened areas, illuminated signs should be provided.

The CAUTION sign is not generally used for area control but should be placed on devices to indicate that they emit MW or RF radiation. Both the WARNING and DANGER signs are used for identifying hazardous devices and for area demarcation.

8.9.8.1 Area control

When the signs are used to designate an exposure area or zone, the following recommendations apply:

(a) the WARNING sign should be placed at the entrance of any zone within which a radiation survey has shown that RF and/or MW power densities exceed the limits recommended. The WARNING sign indicates that LIMITED OCCUPANCY is allowed within its boundaries. It should be located wherever it is necessary to indicate a limit on the occupancy time. In such cases, the WARNING sign should be accompanied by words such as "Warning: Radio-frequency radiation - Maximum occupancy time (t) minutes per hour" where (t) is calculated according to the radiation exposure limits recommended;

(b) the DANGER sign should be placed at the entrance of any zone where a radiation survey has been conducted and where RF and/or MW power densities occur in excess of ten times the continuous occupational exposure limit. The DANGER sign indicates a DENIED

OCCUPANCY zone and should be used in conjunction with barriers that are intended to prevent human access. When used for area demarcation, the DANGER sign should be accompanied by words such as "Danger: Radio-frequency radiation - Do not enter";

(c) where the normal viewing distance of the signs is 1 m or less, the signs should have no outer dimension less than 2 cm. Where the normal viewing distance is more than 1 m, the signs and any associated words should be increased proportionately in size.

Figure 8.1: Examples of warning signs

CAUTION RADIO-FREQUENCY RADIATION

- Negligible hazard
- Minor injury possible from misuse
- Area of unrestricted occupancy

WARNING RADIO-FREQUENCY RADIATION

- Moderate hazard
- Serious injury possible from misuse
- Area of limited occupancy

DANGER RADIO-FREQUENCY RADIATION

- Serious hazard
- Critical injury or death possible from misuse
- Area of denied occupancy

Source: Health and Welfare Canada (12).

8.9.8.2 Labelling devices

When the signs are used for labelling a device, the following recommendations apply:

(a) the WARNING sign should be applied to any device under development or in use for any industrial, scientific or medical purposes, if the device produces exposure levels that exceed the recommended limits. The WARNING sign should be applied to a device if misuse or failure could cause radiation injury;

(b) the DANGER sign should be applied to any device under development or in use for any industrial, scientific or medical purposes, if it produces exposure levels in excess of ten times the appropriate occupational exposure limit.

8.9.9 Special cases

8.9.9.1 Radar and communications systems

The following recommendations apply to radar and communications systems:

(a) where possible, the beam from the antenna should be directed away from occupied areas. If the beam does irradiate occupied areas, the recommended exposure limits should be observed. If needed, antenna sweep restrictions should be employed to reduce the exposure levels;

(b) as previously indicated, excessive RF exposure may occur to maintenance workers repairing waveguides or feed cables to radiating antennae. Workers should be aware of this possibility and take appropriate precautions;

(c) installations to be used for more than one month should have all barriers controlling areas of restricted occupancy and all interlock systems permanently installed;

(d) mobile and transportable systems capable of producing exposure levels exceeding the limits recommended should have warning signs with instructions posted, and clearly visible, regarding the distance at which the radiation levels are below the exposure limits.

8.9.9.2 Dielectric (RF) heaters

Certain types of RF heaters have been found to expose operators to stray RF fields well above 10 mW/cm^2 (equivalent plane-wave power density). Thus special attention is devoted to this type of device. To control RF operator exposure, the procedures given below should be followed (8):

(a) when the RF field intensities are no more than 50 per cent greater than the recommended limits, operator exposure can be controlled by moving the control console for the device further from the applicator. Automatic feeding devices, turntable systems and other means of remote material handling assist in allowing the operators to maintain a reasonable distance from the applicator;

(b) the area of the heater where excessive leakage of RF energy occurs (normally around the applicator) should be shielded or screened. Effective shielding can be provided by thin, flexible metal sheets or straps attached to the upper plate so that they make contact with the lower plate of the applicator when the pressure is applied. The shielding should surround the whole applicator and its effectiveness should be determined by measurements;

(c) a survey of RF fields should be carried out for each heater when it is first installed for use and after any modification that may affect the fields. The survey measurements should be taken with the heater operating with each of its applicators and with the RF output power adjusted to the maximum level specified for each applicator;

(d) the door to the cabinet housing the RF generator should be equipped with at least one interlock. The interlock should prevent generation of RF energy when the door is open;

(e) an indicator that shows when RF power is on should be clearly visible at a distance of at least 3 metres from the heater;

(f) RF heaters should only be used by properly trained personnel. The operators should be aware of precautions to be taken to limit personal exposure. For example, if exposure is controlled by locating the control console at some distance from the applicators, operators should be taught not to approach the device when the RF power is on;

(g) areas in which RF field intensities exceed the recommended limits should be posted and access restricted. The warning signs (see section 8.9.8) should be of such a size that they can be read at a distance of 3 metres;

(h) RF heaters should be maintained and repaired by qualified service personnel, who are aware of all aspects of design, construction and functioning of the device in relation to the control of RF leakage.

8.9.9.3 Microwave and short-wave diathermy

Most applicators used with medical RF diathermy devices are capable of exposing the operators to radiation levels in excess of 10 mW/cm^2. Thus certain precautions are needed to control operator exposure. These include:

(a) examining the diathermy devices regularly (at least annually) to ensure that all parts are in good working order. Particular care should be taken to ensure that cables and electrodes are not cracked or burned, and that the hinges supporting the applicator (electrode) work properly;

(b) testing the operation of electrodes only when they are placed over a tissue-equivalent phantom (e.g. a plastic container of physiological solution);

(c) following the manufacturer's instructions on equipment care. If there is any doubt as to whether the device operates properly, it should be checked by a qualified person;

(d) using diathermy devices only for medically prescribed treatment;

(e) carefully placing the electrodes over the treatment area so as to minimise the gaps between the electrodes and the tissue to be treated. This helps to limit stray RF radiation. Towels may also be used to attenuate stray RF fields in this gap;

(f) turning the RF power on only after the electrodes are in place and the cables supplying power to the applicator(s) are properly connected to the power supply;

(g) the operator remaining at least 1 m from the electrodes and RF cables during the treatment. Moving closer to the electrodes or cables for short periods of time is permitted, but only when necessary;

(h) keeping the area clear of metal beds or chairs during diathermy treatments. Large metal objects should be kept at least 3 m from the electrodes and cables when the diathermy device is in operation (e.g. pillows with metal zip-fasteners should not be used); and

(i) medical diathermy equipment should be used only on patients or test phantoms, never on healthy persons, such as for demonstration or training purposes.

8.9.10 Working procedures and instructions

The occupational health service should, in collaboration with the operator, formulate the radiation protection rules and procedures applicable to the working conditions. These procedures should include the appropriate radiation protection measures to be taken in cases of incidents or accidents. Such procedures should include the transfer of injured personnel to the appropriate medical centres if necessary.

The radiation protection officer should advise the operator of any changes in current working procedures that may be required after evaluating existing practice and any possible need for modifications.

The procedures should include information about the safe working methods and techniques to be followed, and the proper use, operation and care of radiation monitoring instruments and protective devices. They should include the names of the occupational physician and the radiation protection officer. The procedures should be given to workers before taking up their assignments, with special attention to newly recruited and temporary workers.

In addition to the working procedures, all workers should be instructed in their duties and responsibilities before taking up assignments. They should be informed about the sources of exposure to radiation, and the protective measures to be taken and control methods to be followed. Individual radiation workers should, when required, and in addition to the radiation protection rules and procedures, be provided in writing with the necessary detailed working instructions relevant to the specific radiation work to be performed by them.

8.10 Medical surveillance

Medical surveillance or monitoring is established for the following purposes:

(a) the evaluation of workers' physical fitness and condition of health before commencing work and periodic checking of physical fitness during work;

(b) the detection, as early as possible, of clinical signs of any adverse health conditions or possible effects of RF exposure; and

(c) the following exposure history.

Many physical agents produce biological changes in humans which are not considered hazardous to health. Philosophically, one could say that all conditions, including life itself, have profound effects upon biological systems which eventually lead to death. Therefore, we should be practical and define a hazard as an apparent unfavourable effect which leads to some physical injury, disease (physical or mental) or premature death.

In this context a medical surveillance programme should be designed to detect early signs and symptoms of any biological effect before serious health damage has occurred. Unfortunately, in the case of RF radiation, this knowledge is far from complete and thus any present system of health surveillance must be viewed as preliminary in nature.

From a purely medical viewpoint, there may be a limited rationale for the proposal and establishment of a special medical surveillance programme for RF radiation workers at the present time. On the other hand, in the light of the present uncertainty concerning the biological effects, it may be wise (from a legal standpoint) to institute a surveillance programme for the future protection of both the employer and workers. By instituting a health surveillance programme in a large worker population, adequate medical data would become available which could be evaluated for trends. Data obtained through health monitoring could provide insight into potential problem areas that should be subjected to careful epidemiologic study.

8.10.1 Conditions of normal operation

In many countries, initial and periodic medical examinations of workers are a legal requirement. In the USSR, Poland and Czechoslovakia, for example, special medical examinations are mandatory for workers occupationally exposed to RF energy. Such examinations include physical evaluation, haematology and urinalysis, and sometimes a chest X-ray examination. Some countries (e.g. Poland) require more specific examinations (which can include electrocardiography, neurological examination with electroencephalography and opthalmological examination using a slit lamp) for people working with RF energy (33).

In the USSR, Poland and Czechoslovakia, medical examinations may be used to identify people who should not work with RF radiation-emitting devices. Such contra-indications include neoplasia, disorders of the blood-forming system, organic or pronounced functional neurological disturbances and endocrine disorders. In certain cases, such as blood diseases and gastroduodenal ulcer, the worker may be required to work in other areas for the period necessary to regain health before being exposed to RF energy again.

In Western European countries and North America, many industries and government agencies may require a pre-employment medical examination and possibly annual check-ups; however, in most cases no special examinations are performed on RF radiation workers.

When special medical examinations are required for persons occupationally exposed to RF energy, special attention should be paid to define who is exposed. Medical records with information about the exposure history of the workers should be kept by the industrial physician.

8.10.2 Abnormal conditions

When over-exposure occurs, depending on the circumstances, a medical examination may be required. It should be noted that, so far, no unique syndrome requiring highly specialised treatment has been identified as particularly associated with over-exposure to RF radiation. Treatment can be expected to be symptomatic and relatively routine, except that, in an unusual case, one may observe unusually deep burns and tissue necrosis.

In order to ensure that changes in workers' medical histories can be adequately interpreted, account has to be taken of the variations in health, illness and disease patterns that will be encountered in any working population. Therefore, every test or data collected for RF radiation workers must be compared to data collected from large control populations, in order that the statistical nature of the results can be understood. If observations are made on a relatively small group of RF radiation workers for a comparatively small change, especially in young workers, the occurrence of a few cases may be truly uninterpretable.

9. Occupational safety and health services and administrative organisation

Finally, a number of recommendations apply to occupational safety and health services and to organisational arrangements in establishments using radiation sources:

1. The occupational health services serving installations or practices where radiation sources are handled should provide, in addition to their usual tasks, physical and health surveillance to the extent necessary.

2. The functions, organisation and conditions of operation of occupational health services should conform with the requirements laid down in the Occupational Health Services Convention, 1985 (No. 161), and Recommendation, 1985 (No. 171) as adopted by the International Labour Conference.

3. The detailed organisational arrangements needed to ensure adequate radiation protection in a specific installation depend on many factors, including the size of the installation or practice being carried out, and the magnitude of the radiation risk involved. The organisation should be adequate and appropriate to the occupational risks of the installation. Account should be taken of whether the source is confined or unconfined, the nature of the engineering controls available and the extent of supervisory controls present. Attention should be given to the qualifications and competence of the operating personnel to be considered in the organisational plan for radiation protection.

4. Occupational health services should, as far as possible, be located within or near the place of employment, or should be organised in such a way as to ensure that their functions are carried out at the place of employment. The service may be organised as a service within a single enterprise or as a service common to a number of enterprises as appropriate.

5. Occupational health services should have sufficient personnel with specialised training and experience in radiation protection. The service should include (or have arrangements with outside services to have available) a radiation protection officer, a safety engineer and an occupational physician, depending on the size of the installation. The necessary supporting staff for physical surveillance include health physicists and possibly surveyors, as well as the supporting health surveillance services. The heads of the physical surveillance and health surveillance services should have direct access to the employer.

6. The occupational health services should collaborate with those services concerned with the safety of the workers in the enterprise. These include workers' representatives, workers' safety representatives and safety and health committees where these exist.

7. The ultimate responsibility for ensuring adequate protection against radiation rests with the employer and cannot be delegated. However, the employer acting through a well-defined chain of command may, for the purpose of discharging this responsibility, delegate certain of its powers to the staff.

REFERENCES

1. American National Standards Institute (ANSI). 1981. Recommended practice for the measurement of hazardous electromagnetic fields - RF and microwave. 1430 Broadway, New York, NY 10018. ANSI C95.5.

2. ANSI. 1971. Safety guide for the prevention of radio-frequency radiation hazards in the use of electric blasting caps. 1430 Broadway, New York, NY 10018. ANSI C95.4, Pub. No. 20.

3. ANSI. 1973. Techniques and instrumentation for the measurement of potentially hazardous electromagnetic radiation of microwave frequencies. 1430 Broadway, New York, NY 10018. ANSI C95. 3-1973.

4. Assenheim, H.M., Hill, D.A., Preston, E. and Carinie A.B. 1979. The biological effects of radio-frequency and microwave radiation, National Research Council of Canada Report, ISSN 0136-0114, NRCC No. 16448.

5. Baranski, S. and Czerski, P. 1976. Biological effects of microwaves. Stroudburg, Bowden, Hutchinson and Ross.

6. Bassen, H., Kantor, G, Ruggera, P. and Witters, D. 1978. Leakage in the proximity of microwave diathermy applicators used on human and animal models. Rockville, Maryland, United States Department of Health, Education and Welfare. Publication FDA 79-8073.

7. Bureau of Radiological Health (BRH). 1981. An evaluation of radiation emission from video display terminals. Rockville, Maryland, United States Department of Health and Human Services. Publication FDA 81-8153.

8. Canada. 1980. Dielectric (RF) heaters guide-lines for limiting radio-frequency exposure. Ottawa, Health and Welfare Canada. Publication 80-EHD-58.

9. Canada. 1977. Health aspects of radio-frequency and microwave exposure, Part I. Ottawa, Health and Welfare Canada. Publication 77-EHD-13.

10. Canada. 1978. Health aspects of radio-frequency and microwave exposure, Part II. Ottawa, Health and Welfare Canada. Publication 78-EHD-22.

11. Canada. 1979a. "Microwave oven standard", in Canada Gazette (Ottawa, Queens Printer for Canada), 29 Nov., Part 2. P.C. 1979-3259.

12. Canada. 1979. Safety Code-6. Recommended procedures for the installation and use of radio-frequency and microwave devices in the frequency range 10 MHz-300 GHz. Ottawa, Health and Welfare Canada. Publication 79-EHD-30.

13. Commission of the European Communities. 1980. "Proposition de directive du Conseil fixant les principes fondamentaux de la protection sanitaire des travailleurs et de la population contre les dangers des micro-ondes", in Journal officiel des Communuautés Européennes, No. 249/6-9, 26 Sep.

14. Cornelius, W.A. and Viglione, G. 1979. Recommended permissible levels of exposure to microwave and radio-frequency radiation (10 Mhz to 300 GHz) - A proposal, Yallambie, VIC 3085, Australia, Australian Radiation Laboratory. Publication No. ARL/TR 009.

15. Czechoslovakia, Principal Hygienist of CSSR. 1970. "Unified methods for measurement of field intensity and irradiation by electromagnetic waves in the range of high frequency and very high frequency for hygienic purposes", in Priloha c3k Informacnim e pravam 3 oborn hygieny prace a remoci z poovalani. Prague, Institute of Industrial Hygiene and Occupational Diseases, (in Czech), 13 pp.

16. German Democratic Republic, Standard, Industrial Hygiene. 1975. Electric, magnetic and electromagnetic fields and waves - Microwave and high-frequency range definitions, permissible power density, permissible field strength measuring method. Berlin (East). Publication No. DK 628.518:521.3.029.5/6,.

17. Dobrev, B. and Izrael, M. 1981. "Hygienic norms for radio-frequency electromagnetic fields in the People's Republic of Bulgaria", in Health Physics, No. 41, pp. 201-201.

18. Elder R.L., Eure, J.A. and Nicolls, J.W. 1974. "Radiation leakage control of industrial microwave power devices", in Journal of Microwave Power, Vol. 9, No. 2, pp. 51-61.

19. Eure, J.A., Nicolls, J.W. and Elder, R.L. 1972. "Radiation exposure from industrial microwave applications", in The American Journal of Public Health, No. 62, pp. 1573-1577.

20. Federal Republic of Germany. 1979. VDE 0848 Gefahrdung durch electromagnetische Felder, Vol. 1, Bestimmung der Feldgrossen durch Messung und Berechnung. Entwurf, 1 Aug. 1979; Vol. 2, Gefahrdung von Personen durch electromagnetische Felder. Entwurf, 1 Aug. 1979.

21. France. 1981. Sécurité des apareils électrodomestiques et analogues - Deuxième partie: Appareils de chauffage des aliments par micro-ondes. Paris, Association Française de Normalisation (AFNDR). Pub. NF C73-601.

22. France. 1981a. Couleurs et signaux de sécurité. Paris, Norme Française Homologuée NF X 08-003 Additif 1.

23. Fröhlich, H. 1968. "Long-range coherence and energy storage in biological systems", in International Journal of Quantum Chemistry, No. 11, pp. 641-649.

24. Gandhi, O.P. and Chatterjee, I. 1982. "Radio-frequency hazards in the VLF to MF bands", in Proceedings of the IEEE, Vol. 70, No. 12, pp. 1462-1469.

25. Grundler, W. and Keilman, F. 1978. "Non-thermal effects of millimeter microwaves on yeast growth", in Zeitschrift fuer Naturforschung, No. 33C, pp. 15-22.

26. Hankin, N.N. 1979. "Exposure levels and potential biological effects of the PAVE PAWS radar system: Panel on the extent of radiation from the PAVE PAWS radar system", in National Academy of Sciences (Washington, DC), p. 40.

27. Hendler, E. 1968. "Cutaneous receptor response to microwave irradiation", in Hardy, Y.D. (ed.): Thermal problems in aerospace medicine. Surrey, United Kingdom, pp. 149-161.

28. International Electrotechnical Commission (IEC). 1979. Non-ionising radiation hazards in the frequency range from 10 MHz to 300,000 MHz. Geneva, Switzerland. Publication 657, IEC Report.

29. IEC. 1982. Safety in electroheat installations, Part 6: Specifications for safety in industrial microwave heating equipment. Geneva, Switzerland. IEC Publication 519-6.

30. IEC. 1976. Safety of household and similar electrical appliances, Part 2: Particular requirements for microwave cooking appliances. Geneva, Switzerland. IEC Standard, Publication 335-25.

31. IEC. 1978. Safety requirements for radiotransmitting equipment. Geneva, Switzerland. IEC Publication 215, 2nd ed., p. 31.

32. ILO. 1983. Encyclopaedia of occupational health and safety, Vol. II. Geneva.

33. ILO. 1985. Occupational hazards from non-ionising electromagnetic radiation, Occupational Safety and Health Series No. 53. Geneva Switzerland.

34. International Radiation Protection Association (IRPA). 1983. _Interim guide-lines on limits of exposure to electromagnetic fields in the frequency range from 100 kHz to 300 GHz_, International Non-Ionising Radiation Committee (to be published in _Health Physics_).

35. Janes, D.E. 1979. "Radiation surveys - Measurements of leakage emissions and potential exposure fields", in _Bulletin of the New York Academy of Medicine_, Vol. 55, No. 11, pp. 1021-1041.

36. Japan. 1974. _Japanese industrial standard, microwave ovens_. Japanese Standards Association. Tokyo, Japan, Pub. No. JIS C 9250-1974.

37. Lin, R.S., Dischinger, P.C., Conde, J. and Farrell, K.P. 1985. "Occupational exposure to electromagnetic fields and the occurrence of brain tumours", in _Journal of Occupational Medicine_, Vol. 27, No. 6.

38. Logue, J.N., Hamburger, S., Silverman, M.P. and Chiacchierini, R.P. 1985. "Congenital anomalies and paternal occupational exposure to shortwave, microwave, infrared and acoustic radiation", in _Journal of Occupational Medicine_, Vol. 27, No. 6.

39. Michaelson, S.M. 1982. "Microwave radiation", in Suess, M.J. (ed.). _Non-ionising radiation protection_, Copenhagen, WHO. WHO Regional Publications, European Series, No. 10, pp. 97-174.

40. Minin, B.A. 1974. _SVT ch i bezophastnost tcheloveka_. Sovetskow Radio, Moscow, English translation: JPRS Report No. 65506-1, _Microwaves and human safety_. NTIS, Springfield, Virginia, United States.

41. National Council on Radiation Protection and Measurements (NCRP). 1981. _Radio-frequency electromagnetic fields: Properties, quantities and units, biophysical interaction and measurements_. Washington, DC. NCRP Report No. 67.

42. National Institute for Occupational Safety and Health (NIOSH). 1981. _Potential health hazards of video display terminals_. Washington, DC, United States Department of Health and Human Services (DHHS-NIOSH). Publication No. 81-129.

43. Petrov, I.R. 1972. _Influence of microwave radiation on the organism of man and animals_. Springfield, Virginia. NTIS, NASA TT-F.708.

44. Poland, Minister of Health and Social Welfare. 1972. "Decree concerning the definition of electromagnetic fields in the microwave range and the determination of possible occupancy time for work in the hazardous zone", in _Dziennik Urzedowy Ministerstwa Zdrowia i opieki spolecznej_, 78, 20.09, No. 17 (in Polish).

45. Poland, Ministers of Labour, Wages and Social Affairs and of Health and Social Welfare. 1977. "Decree concerning work safety and occupational hygiene during use of equipment generating electromagnetic fields in the range 0.1MHz to 300 MHz", in Dziennik Ustaw, PRL. 19.3, 1977, No. 8 (in Polish).

46. Rabinowitz, J.R. 1973. Possible mechanism for biomolecular absorption of microwave radiation with functional implications. International Microwave Symposium, Abstracts, 314 (IEEE-G-MIT).

47. Repacholi, M.H. 1981. Differentiation between biological effects and health hazards: Scaling for animals. Paper presented at Fourth Course: Advances in biological effects and dosimetry of low-energy electromagnetic fields, Erice, Sicily, 28 March-8 April 1981.

48. Repacholi, M.H. 1982. "Sources and applications of radio-frequency (RF) and microwave energy", in Grandolfo, M., Michaelson, S.M. and Rindi, A. (eds. Biological effects and dosimetry of non-ionising radiation - Radio-frequency and microwave energies. New York. Plenum Press, pp. 19-42.

49. Ruggera, P.S. and Schaubert, D.H. 1982. Concepts and approaches for minimising excessive exposure to electromagnetic radiation from RF sealers. Rockville, Maryland, United States Department of Health and Human Services (DHHS). Pub. (FDA) 82-8192,

50. Ruggera, P.S. 1980. Measurements of emission levels during microwave and shortwave diathermy treatments, Rockville, Maryland, DHHS. Pub. (FDA) 80-8119.

51. Silverman, C. 1980. "Epidemiological studies of microwave effects", in Proceedings of the IEEE, No. 68, pp. 78-84.

52. Smith, A.A. 1978. Attenuation of electric and magnetic fields by buildings, IEEE Trans. Vol. EMC-20, pp. 411-418.

53. Stuchly, M.A., Repacholi, M.H., Lecuyer, D.W. and Mann, R.D. 1982. "Exposure to the operator and patient during shortwave diathermy treatment", in Health Physics, Vol. 42, No. 3, pp. 341-366.

54. Sweden, Worker Protection Authority. 1976. Non-ionising radiation standards, Stockholm.

55. Tell, R.A. 1976. A measurement of RF field intensities in the immediate vicinity of an FM broadcast station antenna. Las Vegas. United States Environmental Protection Agency. Tech. Note OPR/EDA-76-2.

56. Tell, R.A., Hankin, N.N. and Janes, D.E. 1976. "Aircraft radar measurements in the near field, in Carson, P.C., Hendee, W.R. and Hunt, D.C. (eds.). Proceedings of the 9th Midyear Topical Symposium, Boulder, Colorado, Health Physics Society.

57. Tell, R.A. and Mantiply, E.D. 1978. Population exposure to VHF and UHF broadcast radiation in the United States, Las Vegas, NV. USEPA Report No. ORP/EAD 78-5.

58. The Netherlands, Ministry of Health and Environmental Protection. 1979. Hazards of microwave radiation and radiation levels derived therefrom, Report compiled by the Health Council of the Netherlands, No. 66E, Postbox 439-2260AK Leidschendam.

59. United States, ANSI. 1982. Safety levels with respect to human exposure to radio-frequency electromagnetic fields, 300 kHz-100 GHz, ANSI C95, Committee on Radio-Frequency Radiation Hazards, 1430 Broadway, New York. NY 10018.

60. United States. 1970. Performance standard for microwave ovens, Title 42, Pt. 78, Subpt. C., sec. 78.212, Code of Federal Regulation; Federal Register Vol. 35, No. 194, p. 15642, 6 Oct., Washington, DC.

61. USSR. 1976. Electromagnetic fields of radio-frequency: General safety requirements, USSR Standard for Occupational Exposure GOST 12.1.006-76, Occupational Safety Standards System, Moscow, Izdatelstvo, Slandartov, 5 pp. (in Russian).

62. Waker, P.F. and Borman, R.R. 1971. Quantifying hazardous electromagnetic fields: Scientific basis and practical considerations, IEEE Transactions, Microwave Theory and Techniques, MTT-19, p. 178.

63. WHO/UNEP/IRPA. 1981. Criteria I6. Radio-frequency and microwaves. Geneva, WHO.

APPENDIX 1

GLOSSARY

Definitions for this glossary were selected from WHO (63), with permission. Wherever possible, the glossary gives terms and definitions standardised by the International Electrotechnical Commission in the International Electrotechnical Vocabulary (IEV) or by the International Organisation for Standardisation (ISO). In such cases, the IEV number, or the number of the ISO standard in which the definition appears, is given in parentheses.

Absorption - In radio wave propagation, attenuation of a radio wave due to its energy being dissipated, i.e. converted into another form such as heat (IEV 60-20-105).

Absorption cross-section, effective area - Of an antenna, oriented for maximum power absorption unless otherwise stated, an area determined by dividing the maximum power absorbed from a plane wave by the incident power flux density, the load being matched to the antenna (IEV 60-32-035).

Antenna - That part of a radio system which is designed to radiate electromagnetic waves into free space (or to receive them). This does not include the transmission lines or waveguide to the radiator (IEV 60-30-005).

Antenna scanning - See "scanning".

Attenuation - The progressive diminution in space of certain quantities characteristic of a propagation phenomenon (IEV 05-03-115).

Bel; decibel - Transmission units giving the ratio of two powers. The number of bels is equal to the logarithm to the base ten of the power ratio. The decibel is equal to one-tenth of a bel (IEV 55-05-120).

Cavity - A structure that encloses or confines a radio-frequency field.

Conductivity - The scalar or matrix quantity whose product by the electric field strength is the conduction current density (IEV 121-02-1).

Continuous wave - A wage whose successive oscillations are, under steady-state conditions, identical.

Controlled area - An area in which the exposure of personnel to radio-frequency radiation is under the supervision of a safety officer.

Denied occupancy area - An area within which the radio-frequency power density exceeds a set limit.

Dielectric constant, permittivity - A constant giving the influence of an isotropic medium on the forces of attraction or repulsion between electrified bodies (IEV 05-15-120).

Dielectric (material) - A material in which most of the energy required to establish an electric field in the material is recoverable when the field or impressed voltage is removed. A perfect dielectric has zero conductivity and all absorption phenomena are absent. A complete vacuum is the only known perfect dielectric.

Dosimeter - A radiation meter intended to measure the absorbed dose.

Duty factor - The ratio of the sum of pulse durations to a stated averaging time. For repetitive phenomena, the averaging time is the inverse of the pulse repetition rate.

Effective radiated power (in a given direction) - The power supplied to the antenna multiplied by the gain of the antenna in that direction relative to a half-wave dipole (IEV 60-32-095).

Electromagnetic radiation - Radiant energy comprised of varying electric and magnetic fields, which is propagated through space at the speed of light.

Electromagnetic wave - A wave characterised by variations of the electric and magnetic fields (IEV 121-01-38).

Employer - the person responsible for administrative controls over given areas and employees (workers).

Extremities - Limbs of the body including the arms and legs.

Far zone (far field) - See "radiation zone".

Field strength - In radio wave propagation, the magnitude of a component of specified polarisation of the electric or magnetic field. The term normally refers to the root-mean-square value of the electric field (IEV 60-20-070).

Free space - An ideal, perfectly homogeneous medium possessing a dielectric constant of unity and in which there is nothing to reflect, refract or absorb energy. A perfect vacuum possesses these qualities.

Frequency - The number of cycles per second; the reciprocal of period.

Fresnel region - Of a transmitting antenna system, the region near the antenna system where the wavelets arriving from the various parts of the system cannot be considered to follow parallel paths (IEV 60-32-065).

Induction field - That part of the field of an antenna which is associated with a pulsation of energy to and fro between the antenna and the medium. (Note: The induction field extends theoretically over the whole of space, but is negligible compared with the radiation field except in the neighbourhood of the antenna (IEV 60-32-045).)

Induction zone; near zone - The region surrounding a transmitting antenna in which there is a significant pulsation of energy to and fro between the antenna and the medium. (Note: The magnetic field strength (multiplied by the impedance of space) and the electric field strength are unequal and, at distances less than one-tenth of a wavelength from an antenna, vary inversely as the square or cube or the distance, if the antenna is small compared with this distance (IEV 60-32-055).)

Interlock - A device that prevents the activation of a given mechanism or process unless certain precautionary measures have been observed.

Irradiation, partial body - Exposure of only part of the body to incident radio-frequency energy.

Irradiation, whole body - Exposure of the entire body to incident radio-frequency energy.

Isotropic - Having the same properties in all directions.

Leakage radiation - Unintended, unwanted or non-useful radiation transmitted by a device into accessible areas.

Limited occupancy area - An area within which the radio-frequency power density is greater than that specified for unlimited occupancy but less than a set limit.

Microwaves - Electromagnetic waves of sufficiently short wavelength that practical use can be made of waveguide and associated cavity techniques in their transmission and reception (IEV 60-02-025). (Note: for the purposes of this review the term is taken to signify waves having a frequency range of 0.3-300 GHz).

Modulated wave (radiation) - When a continuous series of waves of electromagnetic energy is modified by pulsing, or by varying its amplitude, frequency or phase, the waves are called, respectively, pulse-, amplitude-, frequency-, or phase-modulated.

Near zone (near field) - See "induction zone".

Occupancy time - The time period (in minutes) that a radiation worker is allowed to spend in a limited occupancy area within the averaging time of the exposure standard (e.g. 6 min in the case of the United States, 1982).

Owner - A person, organisation or institution having title to administrative control over a given device emitting RF radiation. The owner is the owner of the source of RF radiation and is responsible to all people in the facility, not just the employees.

Polarisation, plane of - In a linearly polarised wave, the fixed plane parallel to the direction of polarisation and the direction or propagation.

Power - (1) Mean power, work (or energy) divided by the time in which this work (or energy) was produced or absorbed. In periodic phenomena, the average power during a period is generally taken. (2) Instantaneous power, the limit of the average power when the interval of time considered becomes infinitely small (IEV 05-04-025).

Power flux density; field intensity - In radio wave propagation, the power crossing unit area normal to the direction of wave propagation (IEV 60-20-075).

Pulse duration - The interval of time between the first and last instant at which the instantaneous value of a pulse (or of its envelope if a carrier frequency pulse is concerned) reaches a specified fraction of the peak amplitude (IEV 55-35-105).

Pulse repetition rate - The average number of pulses in unit time during a specified period (IEV 55-35-125).

Radar - An acronym for radio detection and ranging. The use of radio waves, reflected or automatically retransmitted, to gain information concerning a distant object. The measurement of range is usually included.

Radar scan - See "scanning".

Radiation field - That part of the field of an antenna which is associated with an outward flow of energy (IEV 60-32-040).

Radiation survey - An evaluation of the actual or potential (due to malfunction) radiation levels in any area, including a review of pertinent records, specifically in the vicinity of radio-frequency devices.

Radiation zone; far zone - The region sufficiently remote from a transmitting antenna for the energy in the wave to be considered as outward flowing. (Note: In free space, the magnetic field strength (multiplied by the impedance of space) and the electric field strength are equal in this region and, beyond the Fresnel region, vary inversely with distance from the antenna. The inner boundary of the radiation zone can be taken as one wavelength from the antenna, if the antenna is small compared with this distance (IEV 60-32-050).)

Radio-frequency - Any frequency at which electromagnetic radiation is useful for telecommunication. In this review, this term is used to designate frequencies ranging from 100 kHz to 300 GHz.

Responsible user - A person designated by the owner to have immediate authority for the use and operation of a given radio-frequency device.

RF worker - A person who is employed to work with radio-frequency devices.

Safety officer - A person appointed by the owner to evaluate safety techniques and provide guidance regarding RF radiation protection.

Scanning - Of a radar antenna, systematic variation of the beam direction for search or angle tracking (IEV 60-72-095). The term is also applied to periodic motion of a radio communication antenna.

Shield - A mechanical barrier or enclosure provided for protection (IEV 151-01-18). The term is modified in accordance with the type of protection afforded, e.g. a magnetic shield is a shield designed to afford protection against magnetic fields.

Specific absorption rate (SAR) - The rate at which electromagnetic energy is absorbed at a point in a medium per unit mass of the medium (normally expressed in W/kg). SAR is analogous to the "absorbed dose rate" used in the ionising radiation field.

Uncontrolled area - Any area that is not under the supervision of a safety officer.

Wave - A modification of the physical state of a medium which is propagated as a result of a local disturbance (IEV 05-03-005).

Wave, plane - A wave such that the corresponding physical quantities are uniform in any plane perpendicular to a fixed direction (IEV 05-03-010).

Waveguide - A system for the transmission of electromagnetic energy by a wave not of TEM type. It may, for example, consist of a metal tube, a dielectric rod or tube, or a single wire (IEV 62-10-005).

Wavelength - The distance between two successive points of a periodic wave in the direction of propagation, in which the oscillation has the same phase (IEV 05-03-030). Symbol: λ.

X-radiation - Electromagnetic radiation having a frequency greater than 3×10^{16} Hz, i.e. wavelength in free space of less than 10 mm.

APPENDIX 2

PHYSICAL PROPERTIES OF RADIATION

Figure 1: A monochromatic electromagnetic wave

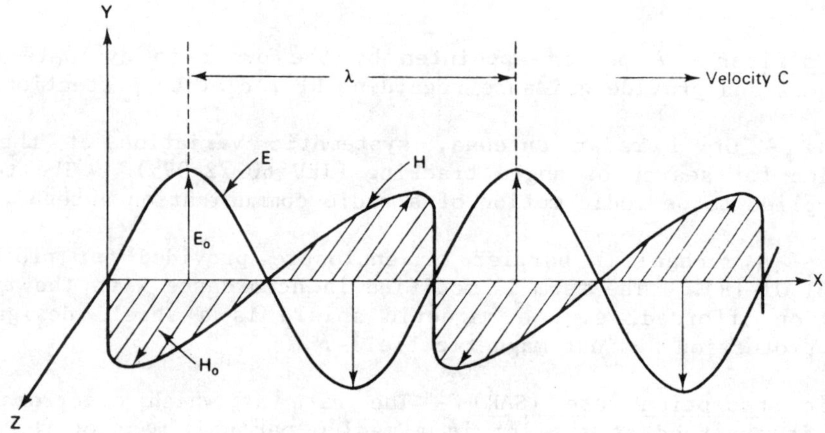

In free space, electromagnetic waves spread in all directions from a point source (figure 1). The wavefront - the plane joining all points of identical phase - is spherical in this case. As the distance from the point source increases, the area of the wavefront surface increases as the square of the distance. In this case the power density W is defined as the ratio of the radiated power per unit area of a surface perpendicular to the incident wave. The power density is inversely proportional to the square of the distance from the source and can be expressed as:

$$W = \frac{P}{4\pi r^2}$$

where P is the transmitted power and r is the distance from the source.

An isotropic source radiates uniformly in all directions in space; in practice, there is no source having this property, although there are close approximations. The inverse-square law also applies when the source is not isotropic but the medium is isotropic, e.g. free space, that is when the velocity of wave propagation does not change with the distance.

An electromagnetic wave can also be characterised by the electric field strength and magnetic field strength. The electric field strength at a distance r from a point source radiating power P is equal to:

$$E = \sqrt{ZP/4\pi r^2}$$

and is expressed in volts per metre (V/M). In the far field the relationship between electric field strength and power density W (normally expressed in units of W/m^2 or mW/cm^2 is:

$$W = \frac{E^2}{Z}$$

where Z is the characteristic impedance of the medium, for free space $Z = 377\Omega$. The magnetic field strength is related to the power density by the following relation:

$$W = H^2 Z$$

H is expressed in amperes per metre (A/m). The characteristic impedance is defined by the permittivity and permeability of the medium where:

$$Z = \sqrt{\mu/\varepsilon}$$

with $= 8.85 \times 10^{-12}$ F/m and $= 1.26 \times 10^{-6}$ H/m for free space.

Electromagnetic waves are reflected from metal surfaces. In general, the reflection coefficient, defined as the ratio of the electric field strength of the reflected wave to that of the incident wave, is equal to 1 for a perfect conductor and very close to 1 for practical conducting surfaces. The behaviour of electromagnetic waves incident upon a non-metallic surface and their propagation within a dielectric material is defined by the dielectric permittivity and magnetic permeability of the material. Magnetic permeability of biological materials is the same as that of free space. The dielectric permittivity is a complex variable:

$$\varepsilon = \varepsilon' - j\varepsilon''$$

where ε' is the dielectric constant and ε'' is the loss factor. The dielectric constant is a measure of the medium's ability to store electric energy and defines the strength of the electric field within the material exposed to a microwave field. The loss factor describes what fraction of the energy is dissipated in the material. The loss tangent, sometimes called the dissipation factor, is defined as:

$$\tan \delta = \varepsilon''/\varepsilon'.$$

The permittivity of materials depends on their physical state (solid, liquid or gas), composition (content of water, salts, types of solids present, etc.), temperature and frequency.

When electromagnetic waves pass from one medium to another having different dielectric properties, part is reflected and the remaining part propagates into the second medium, frequently with a different direction of propagation. The propagation within the second medium depends on the medium's permittivity and permeability. Figure 2 shows a plane wave at an interface between two media. If the second medium is lossy, the wave is attenuated and the strengths of the electric and magnetic fields decrease. The distance at which the amplitude of the field decreasses to 1/e of the amplitude at the surface is called the penetration depth or skin depth.

The field produced by an antenna can be divided into two regions: near-field zone (Fresnel region), and far-field zone (Fraunhofer region). The criterion most frequently used to define the division of the space between the near and far fields is that the phase of the fields from the points of the radiating antenna aperture does not differ more than the equivalent of /16, where is the wavelength of the radiation. As shown in figure 3, if S is to be less than or equal to /16, the minimum distance R where the far-field zone begins can be found to be equal to:

$$R = \frac{2a^2}{\lambda}$$

where a is the longest dimension of the antenna. In the far-field zone the antenna radiation pattern is practically independent of the distance from the antenna, while in the near field the pattern is, in general, a function of the distance at which it is measured.

Figure 2: A wave incident on a lossy dielectric material

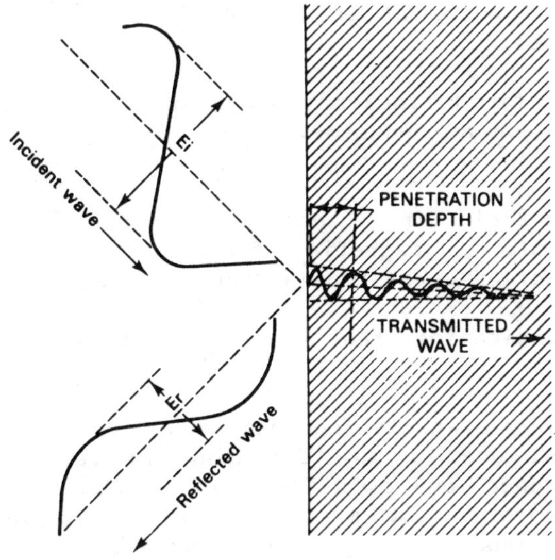

Figure 3: Definition of the antenna far-field zone

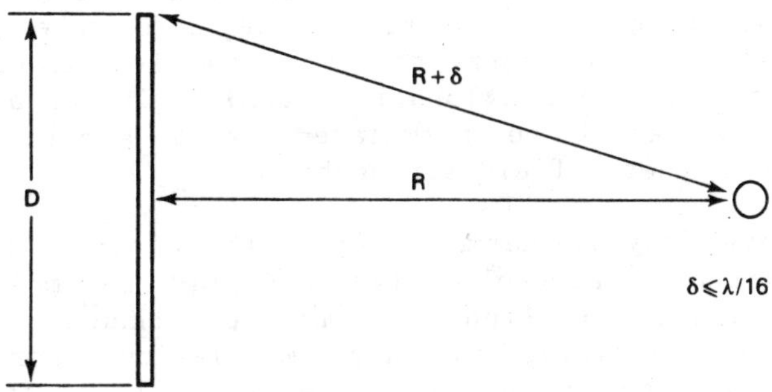

The basic characteristic of electromagnetic waves may be summarised as follows:

1. Time-varying electrical currents are the source of electromagnetic waves.

2. The electromagnetic wave may be represented as a vector field consisting of two components, the electric vector and magnetic vector.

3. Propagation of fields from the source is radial and has a wave nature; a change in the medium of propagation affects this

phenomenon according to the dielectric constant, conductivity and permeability of the medium.

4. Partial transmission and partial reflection of the electromagnetic wave occur at the boundary of media having different electrical properties.

5. Electric field lines may begin and end on charges; if an electric field ends on a conductor, it must represent a charge induced on a conductor.

6. Magnetic field lines cannot end; since no magnetic charges are known physically, magnetic fields form continuous loops surrounding a conduction current or a changing electric field.

7. Lines indicating the direction of propagation may be referred to as rays.

Using the power density concept and considering MW and RF radiations as rays, one is apt to think in terms of geometrical optics. All the phenomena of diffraction, interference, reflection, dispersion and absorption of microwaves exist in various media. Geometrical optics must be replaced, however, by physical optics, because microwaves are much longer than visible light rays and are comparable to the dimensions of the irradiated objects. This sort of optics may be applied only in the far-field zone; in the near-field zone it may be inapplicable. For example, a standing wave formed by two single plane waves of equal amplitude, same polarisation and opposite directions has a time-average zero power density, but the electric and magnetic energy densities may be quadruple those of the original waves at certain locations (62). In the near-field zone the power density cannot be formulated as a simple function of the electric and magnetic field strengths.

The waves may be harmonic (i.e. the electric and magnetic field vectors oscillate sinusolidally) and may be modulated (i.e. the amplitude, phase or frequency may be changed in an arbitrary manner). For example, in pulse modulation short electromagnetic pulses are sent out at certain time intervals. The duration of the pulse (pulse length, pulse width) is usually of the order of milliseconds. The pulse repetition rate is expressed in hertz or cycles per second.

The product of pulse length and repetition rate is referred to as the duty cycle. In the case of pulsed-wave generation, the power emitted increases rapidly, reaches a peak pulse power and rapidly decreases. The peak pulse power divided by the duty cycle gives the average power emitted (for rectangular pulse waveform). It should be realised that the peak pulse power is many times higher than the average power output. For example, the latter may be of the order of kilowatts, whereas peak power may be of the order of megawatts.

Source: ILO (33, pp. 4-9).

APPENDIX 3

SURVEY FORMS

There is certain basic information that is required and must be obtained when conducting radiation safety surveys. This information is itemised below. In addition, specialised information is needed depending on the type of survey being conducted, e.g. device, facility, installation of a number of devices. Examples of this specialised information are also given below.

Basic survey information

1. Inspector (name)
2. Date of inspection
3. Survey location
4. Names of owner and persons contacted
5. Device(s): Manufacturer, model, serial number, date of manufacture
6. Device(s) operating parameters: Frequency (nominal operating), output power, CW or pulsed, pulse repetition frequency, pulse duration, normal operation time per day
7. Operator(s) location(s), occupancy time (per working day or per hour)
8. Radiation measurements

Examples of specialised survey information
(e.g. microwave heating installation)

1. Indicator for status of power?
2. Interlocks on access doors? Locations.
3. Warning signs: Description, location.
4. Test load description - when used during radiation measurement.

APPENDIX 4

X-RADIATION EMISSION

High-power RF microwave devices often require high operating voltages, typically above 15-20 kV, and thus there is the likelihood of X-ray production. When the manufacturer's data or other observations indicate that the generation of X-rays is possible, the following procedures and exposure limits are recommended:

(a) the RF or MW power-generating component should be inspected to determine if specified X-ray shields are in their designated locations;

(b) with the shields located correctly, an X-radiation survey should be conducted to ensure that the intensity of X-rays, averaged over an area of 10 cm^2 at 5 cm from the external surface of the device, does not exceed 0.5 mR/min;

(c) survey instruments used for the measurement of X-radiation intensity must not be susceptible to RF radiation. In selecting instruments for measuring X-rays from pulsed sources such as magnetrons and klystrons, only those capable of making measurements, where instantaneous exposure rates may be high, should be chosen. Many ion chambers and most Geiger-Muller counters are unsuitable because of saturation effects. Integrating dosimeters such as LiF work well;

(d) survey procedures, reports and records should be similar to those in section 8.8 of this document.